DRAWING TO LEARN BIOLOGY

Students Assignments in Illustrating Biology Concepts

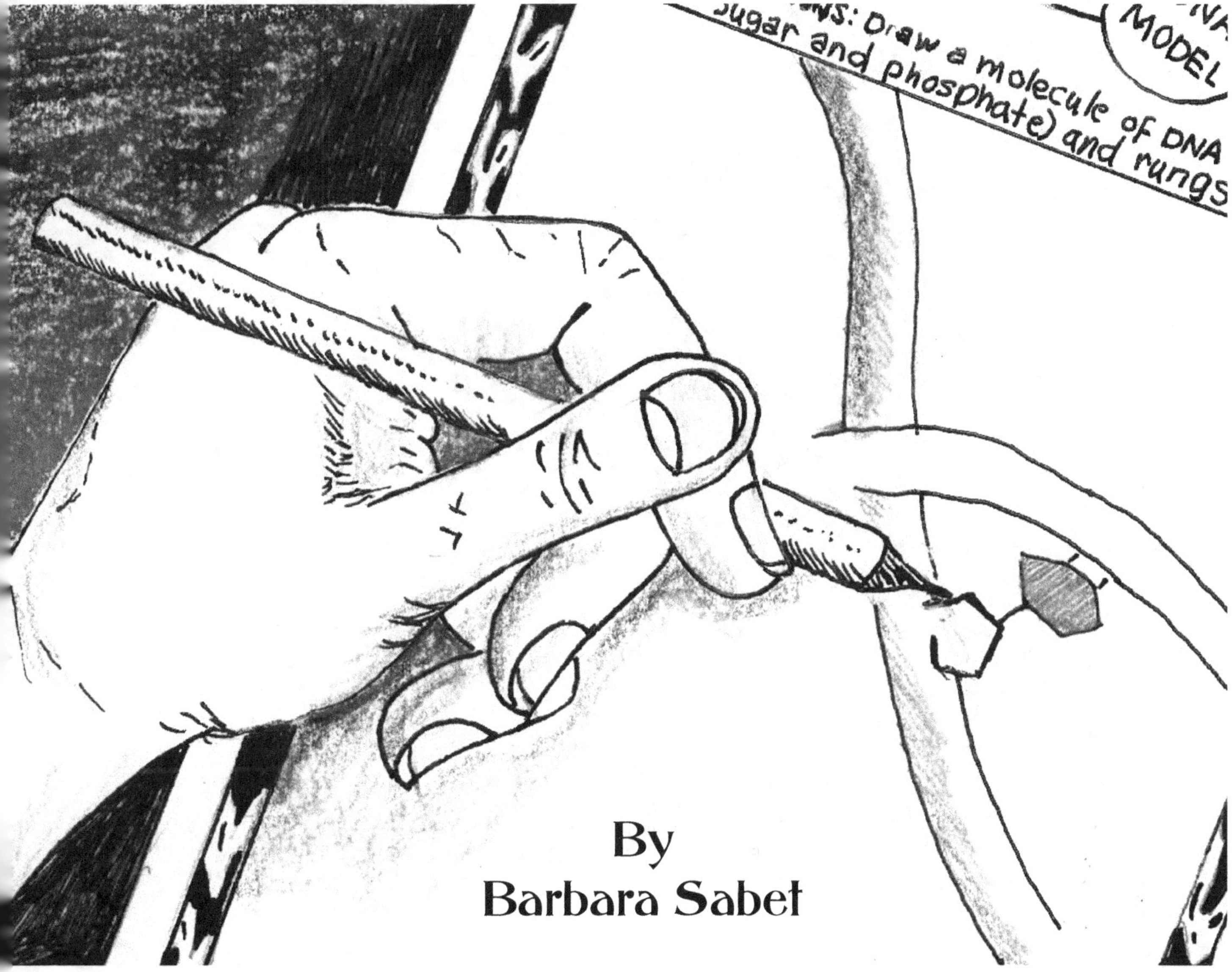

By
Barbara Sabet

Copyright 2016
All rights reserved
ISBN: 13: 978-1539983170

Table of Contents

1. Art Forms in Nature — page 5
2. The 6 Kingdoms of Life — page 7
3. Enzyme Action — page 9
4. Elements in Organisms — page 11
5. Macromolecules in Organisms — page 13
6. Heirarchy of Life — page 15
7. Cells Analogy — page 17
8. Cell Types — page 19
9. Homeostasis — page 21
10. The Cell Membrane — page 23
11. ATP — page 25
12. Life Cycle — page 27
13. Photosynthesis — page 29
14. Cellular Respiration — page 31
15. The Chromosome — page 33
16. Cell Cycle — page 35
17. Growth and Development Timeline — page 37
18. Mitosis — page 39
19. Meiosis — page 41
20. DNA Model — page 43
21. DNA Replication — page 45
22. RNA — page 47
23. Protein Synthesis — page 49
24. DNA Fingerprinting — page 51
25. Recombinant DNA — page 53
26. Human Mutations — page 55
27. Tree of Life — page 57
28. Evidence for Evolution — page 59
29. Natural Selection — page 61
30. Adaptation — page 63
31. Ecology in Levels — page 65
32. Species Interactions — page 67
33. Food Web — page 69
34. Bio-magnification — page 71
35. Biogeochemical Cycles — page 73
36. Greenhouse Effect — page 75
37. Reduce, Re use, and Recycle — page 77
38. Human Body Anatomy — page 79
39. Nervous System — page 81
40. Endocrine System — page 83
41. Immune System — page 85
42. Integumentary System — page 87
43. Digestive System — page 89
44. Choose a Body Organ — page 91
45. Anatomy of the Frog — page 93
46. Sample pages — page 95-139

Name: _____ Period: _____ Date: _____

Biology Drawing — Art Forms in Nature

Directions: Find a natural object, such as a leaf or insect. Draw it and color it in an artistic way to showcase art in nature as done by many scientists in the past

Name: _____ Period: _____ Date: _____

Biology Drawing — The 6 Kingdoms of Life

Directions: Divide the space below into 6 sections. Label each section with one of the 6 kingdoms. Draw a representative of each kingdom, name it, and write down the characteristics of each kingdom in the space. You may add cut-out drawings to make it look like a collage if your teacher so desires.

Name: _____ Period: _____ Date: _____

Biology Drawing — Enzyme Action

Directions: Draw an enzyme showing the active site and then use the molecules shown to fit into the active site. The enzyme can be used to bring the two molecules together or to split one molecule into two. Label.

Name: _____ Period: _____ Date: _____

Biology Drawing — Elements In Organisms

Directions: Divide the space below into 4 sections where you will draw a carbon, hydrogen, nitrogen and oxygen atoms which constitute 95% of your body weight. Be sure to show the nucleus, orbitals, electrons, protons & neutrons. Label and color. Use the circle below in your picture.

Name: _____ Period: _____ Date: _____

Biology Drawing — Macromolecules in Organisms

Directions: Divide the space into 4 sections and draw the monomers of each of the 4 macromolecules (carbohydrate, protein, fat and nucleic acid). Use the same color for the atoms. i.e. Carbon could be black, oxygen red and hydrogen white. Use the symbol below for the monomer glucose.

Name:_____ Period:____ Date:_____

Biology Drawing — Heirarchy of Life

Directions: Draw a way to show the heirarchy of life from **organelle** to **cell** to **tissue** to **organ** to **organ system** and finally, to **body**.

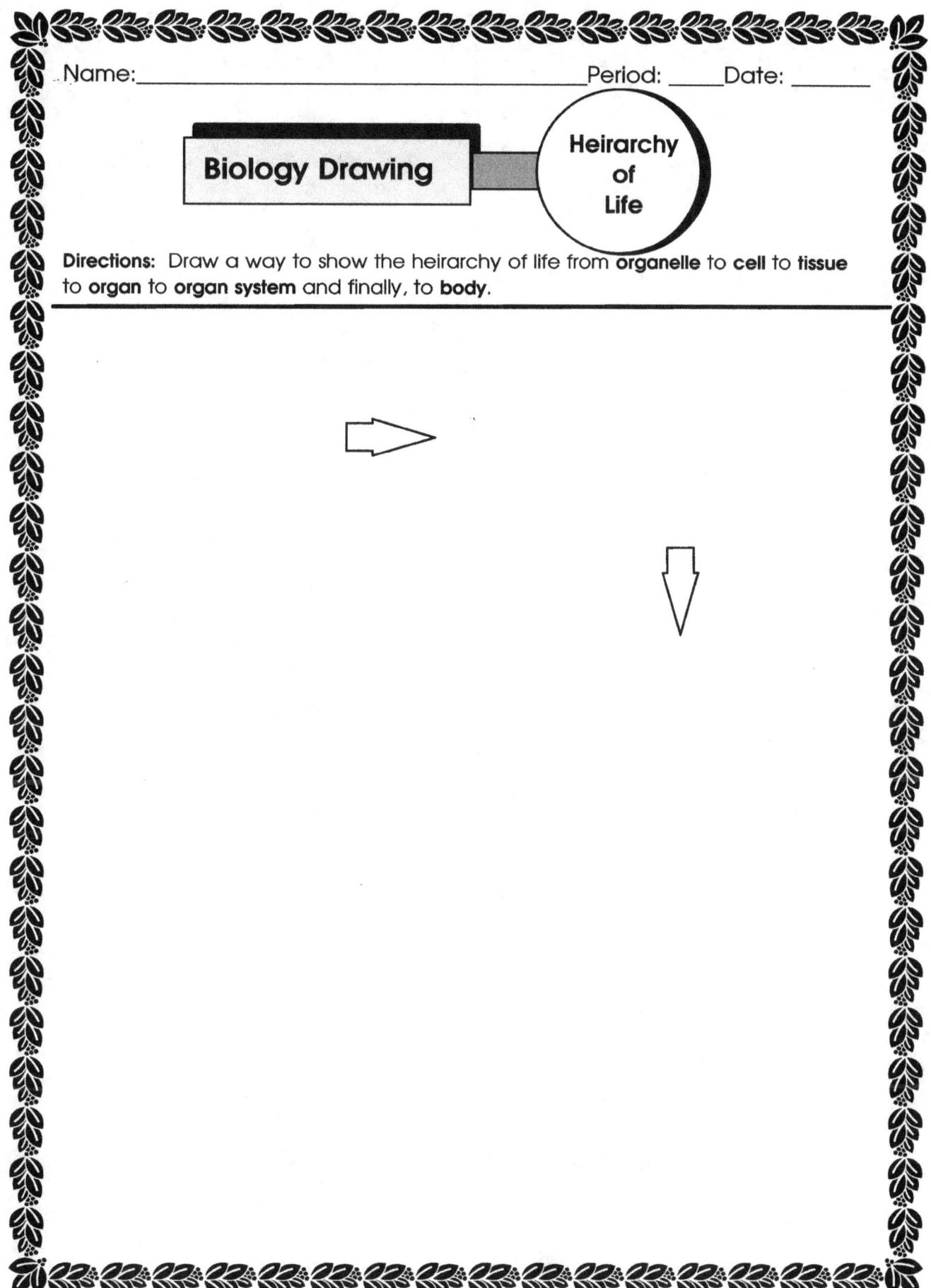

Name: _____ Period: _____ Date: _____

Biology Drawing — Cell Analogy

Directions: The cell is often compared to towns, schools, factories and other common systems. Pick one of these or something else to compare the cell to. Divide the space in half, showing the cell in one half, and the analogy in the other, using the same colors for comparison. i.e. the mitochondria and the town's powerplant would both be colored brown. Be sure to label. Use the symbol below in your drawings.

Name: _____ Period: _____ Date: _____

Biology Drawing — Cell Types

Directions: Divide the space into 4 sections. Label the sections: animal cell, plant cell, protist cell, and bacterial cell (prokaryotic). The first three cells are eukaryotic. Draw a general representative of each type of cell, and then color code the drawings so that the same organelles have the same colors throughout. Use the symbol below in your drawing.

Name:_____ Period:_____ Date:_____

Biology Drawing — Homeostasis

Directions: Draw a representation of a negative feedback loop, showing a real-life example. Don't forget to label all the steps in the circular loop below.

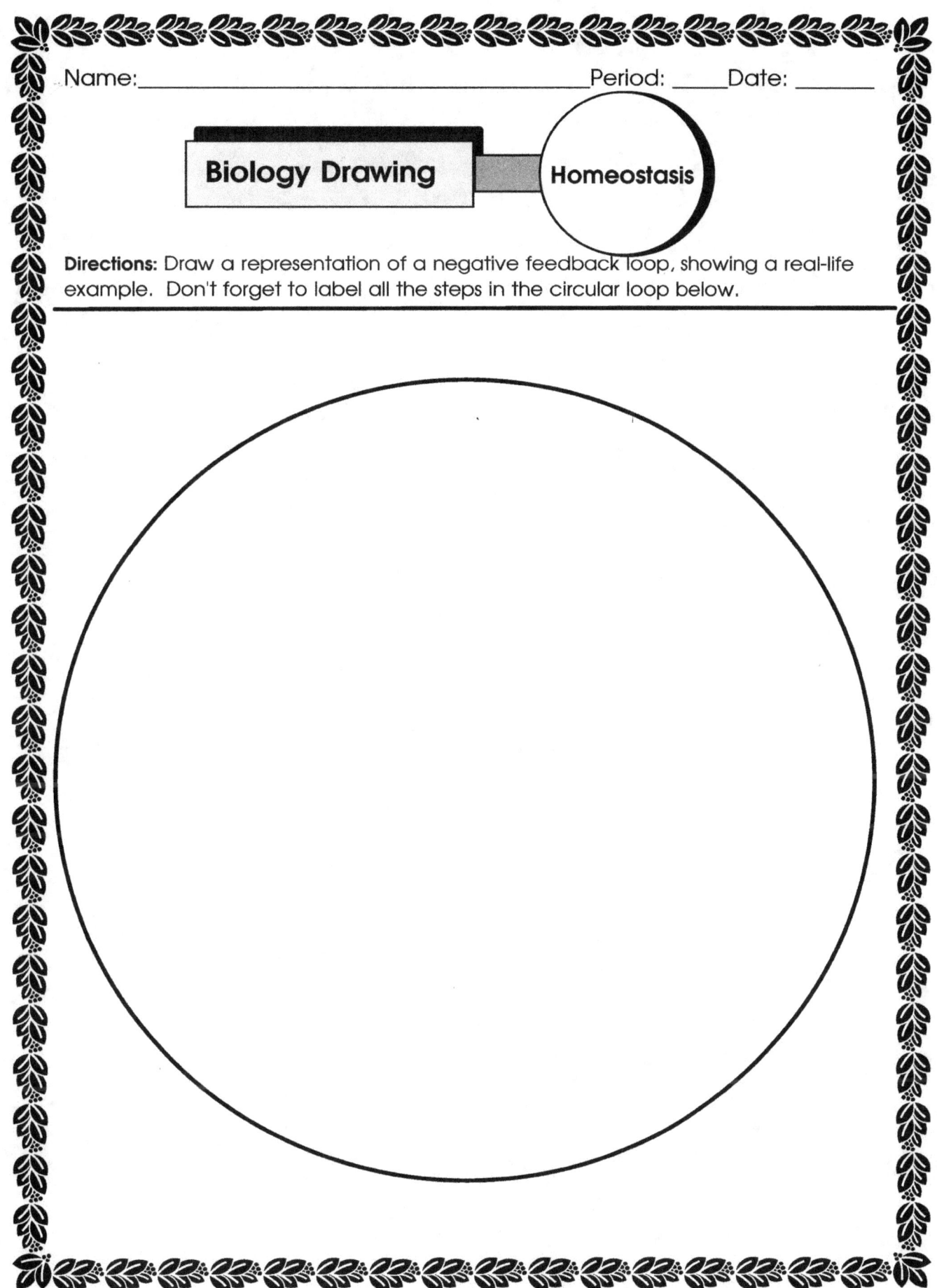

Name: _____   Period: _____ Date: _____

Biology Drawing — The Cell Membrane

Directions: Draw a cell membrane showing the different types of proteins embedded in the phospholipids. Then, show movement through the membrane.

Name:_____ Period:____ Date:_____

Biology Drawing — ATP

Directions: Draw the structure of ATP and show how it stores energy within it's high energy bonds.

Name: _____ Period: _____ Date: _____

Biology Drawing — Life Cycle

Directions: Using the circle below, draw the components of the life cycle, showing how animals and plants keep each other alive with the exchange of gases and other molecules.

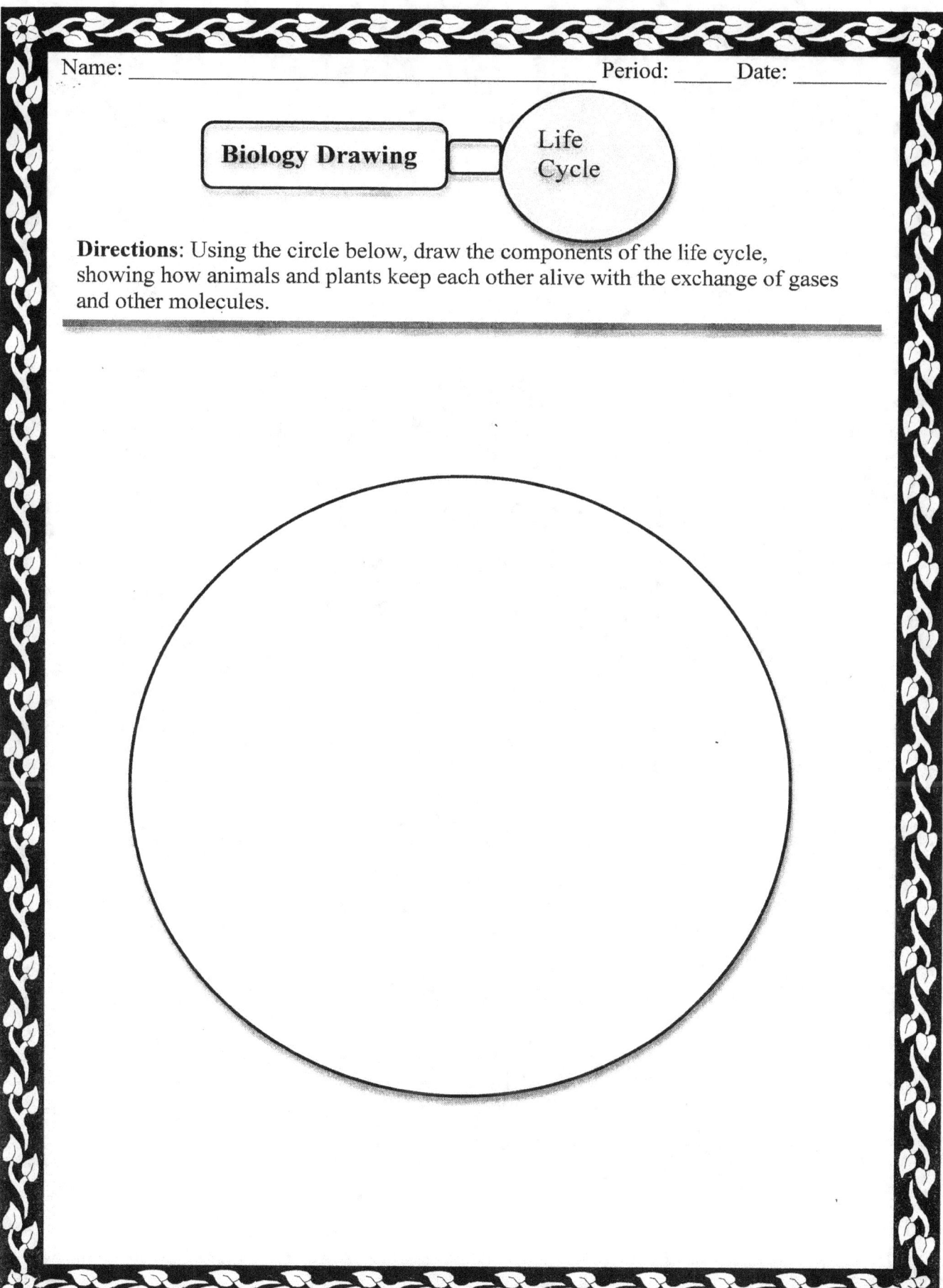

27

Name: _____ Period: _____ Date: _____

Biology Drawing — Photo-synthesis

Directions: Draw a simplified view of photosynthesis, giving as much detail as your teacher requests. Be sure to label and color, using the sun below.

29

Name: _____ Period: _____ Date: _____

Biology Drawing — Cellular Respiration

Directions: Draw a simplified view of cellular respiration, giving as much detail as your teacher requests. Be sure to label and color, using the circle below for the Kreb's cycle

Name: _____ Period: _____ Date: _____

Biology Drawing — The Chromosome

Directions: Draw a chromosome below with as much detail as you can. If directed by your teacher, draw the DNA inside the chromosome and show how it wraps around the histone proteins. Label, color and use the symbol below.

Name: _____ Period: _____ Date: _____

Biology Drawing — The Cell Cycle

Directions: Use the circle below to draw the cell cycle. Use a ruler to divide it into the phases of the cycle. Label the phases and draw what the chromosome looks like in each phase and tell what is happening in each phase.

Name:_____ Period: _____ Date: _____

Biology Drawing — Growth and Development Timeline

Directions: Draw a timeline of the growth and development of the human starting with the fertilized egg and ending with the birth of the baby. Be sure to show the morula, blastocyst, fetus, embryo, etc.

Name: _____   Period: _____ Date: _____

Biology Drawing — Mitosis

Directions: Draw a cell undergoing mitosis, showing all of the steps and labeling each. Use at least 4 chromosomes in the drawing (2 homologous pairs). Use different colors for each pair. Show spindle fibers also. Use the symbol below.

Name: _____ Period: _____ Date: _____

Biology Drawing — Mieosis

Directions: Draw a cell undergoing meiosis, showing all of the steps and labeling each. Use at least 4 chromosomes in the drawing (2 homologous pairs). Use different colors for each pair. Show spindle fibers also. Use the symbol below.

Name: _____ Period: _____ Date: _____

Biology Drawing — DNA Model

Directions: Draw a molecule of DNA at least 2 codons long showing the sides sugar and phosphate) and rungs (nitrogen bases) of the ladder.

Name: _____ Period: _____ Date: _____

Biology Drawing — DNA Replication

Directions: Show how DNA replicates within the nucleus. Use the amount of detail that your teacher requests. Use the symbol provided (can be a nitrogen base).

45

Name: _____ Period: _____ Date: _____

Biology Drawing — RNA

Directions: Show the three kinds of RNA in the space below: Transfer RNA, Messenger RNA and Ribosomal RNA. Use the symbol below in your drawing.

47

Name: _____ Period: _____ Date: _____

Biology Drawing — Protein synthesis

Directions: Draw a diagram showing both of the major steps in protein synthesis: transcription and translation. Use the ribosome below in your drawing

49

Name: _____ Period: _____ Date: _____

Biology Drawing — DNA Finger-printing

Directions: Draw the process of DNA fingerprinting, showing the electrophoresis part of the process as well. Label, color and include the symbol below.

51

Name: _____ Period: _____ Date: _____

Biology Drawing — Recombinant DNA

Directions: Draw the process whereby recombinant DNA is utilized to make important biological molecules. Be sure to include the plasmid, chimera and donor DNA. Color, label and include the figure below.

53

Name:_____ Period:_____Date:_____

Biology Drawing — Human Mutations

Directions: Draw a representation of the types of mutations with examples. It can be in a chart form if you like. Include gene mutations and chromosome mutations.

55

Name:_____ Period:_____ Date:_____

Biology Drawing — Tree of Life

Directions: Draw a tree-like structure depicting the tree of life with the 3 domains **Bacteria**, **Eukarya** and **Archaea** as the 3 main branches.

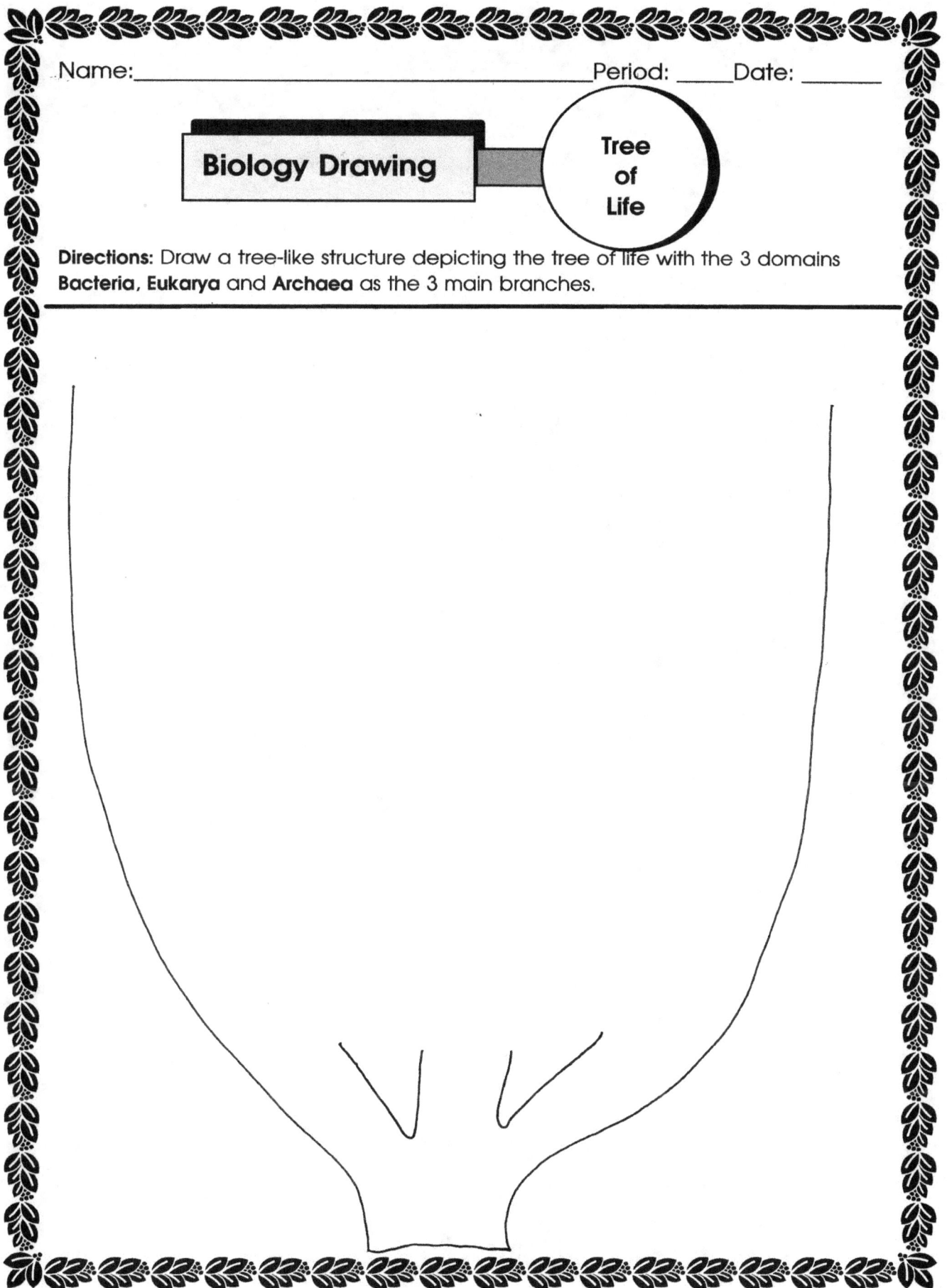

Name: _____ Period: _____ Date: _____

Biology Drawing — Evidence for Evolution

Directions: Draw a representation of the 5 evidence categories for Evolution, such as fossil evidence, comparative anatomy, comparative embryology, comparative biochemistry and biogeography. You can make a chart if you like.

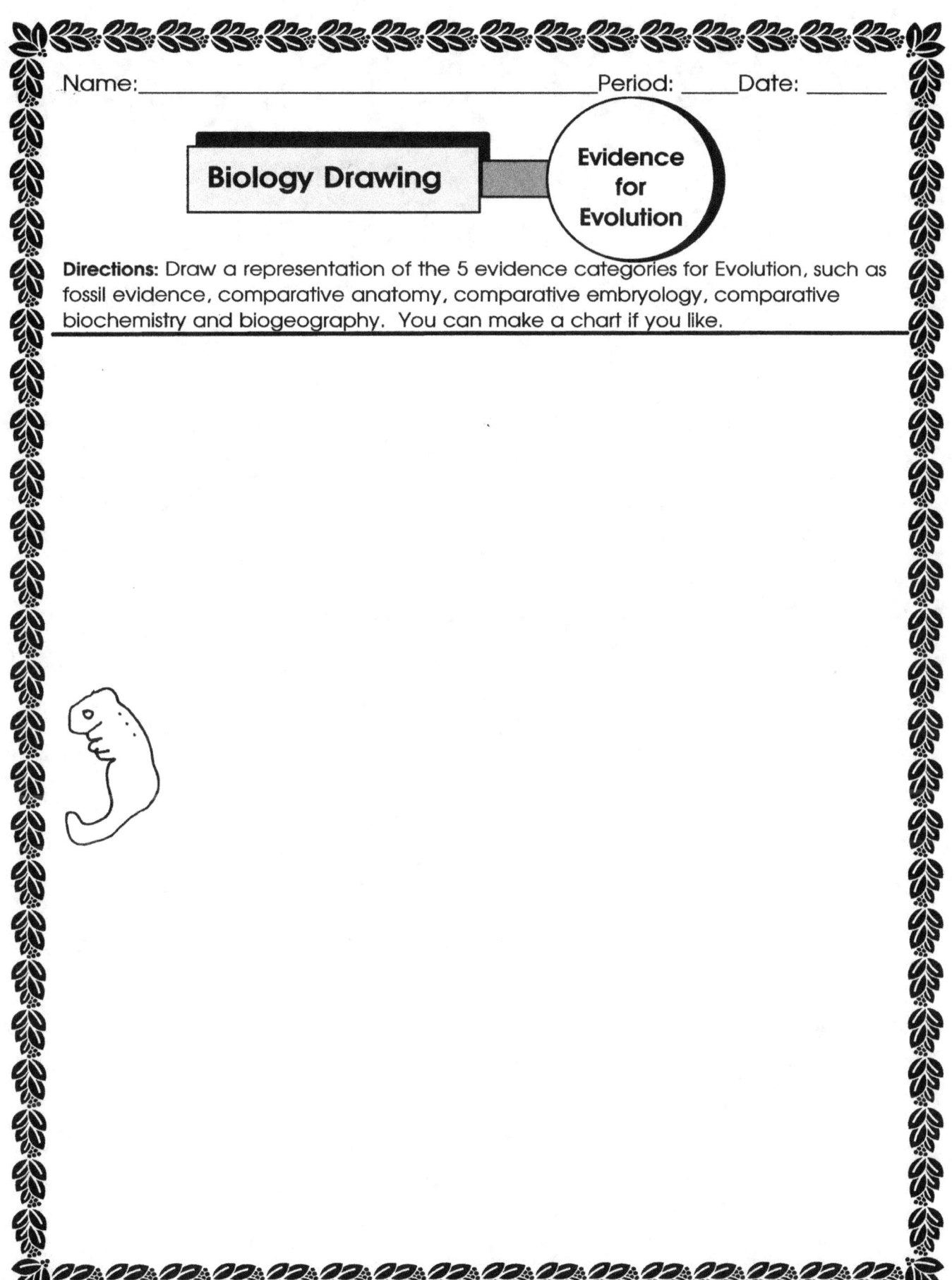

Name: _____ Period: _____ Date: _____

Biology Drawing — Natural Selection

Directions: Divide the page into 5 boxes and show the 5 steps for natural selection as told to you by your teacher. Use the line below to guide you.

61

Name: _____ Period: _____ Date: _____

Biology Drawing — Adaptation

Directions: Adaptations help organisms to survive and/or reproduce. They can be structural, physiological or behavioral. Pick an animal with a particular adaptation and draw him showing that adaptation in the middle box. In the first box, show what that animal may have looked like millions of years ago before he developed the adaptation. In the last box, show what the animal may look like in a million years with a new adaptation or a change to the original adaptation. Explain what happened on the back and include the symbol below.

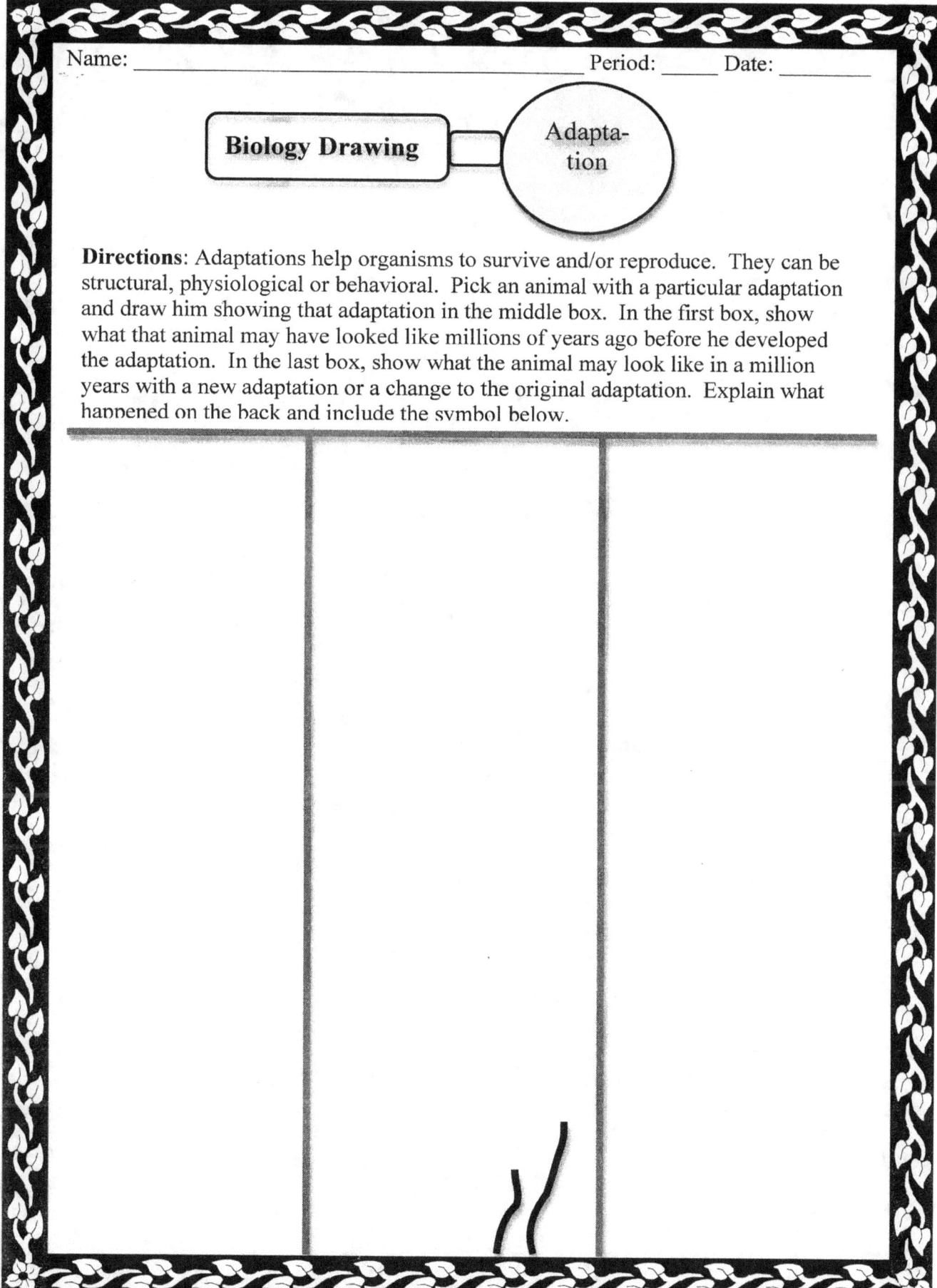

Name: _____ Period: _____ Date: _____

Biology Drawing — Ecology in Levels

Directions: Choose an organism, and then show how it fits into a population, community and ecosystem. Use the circles below.

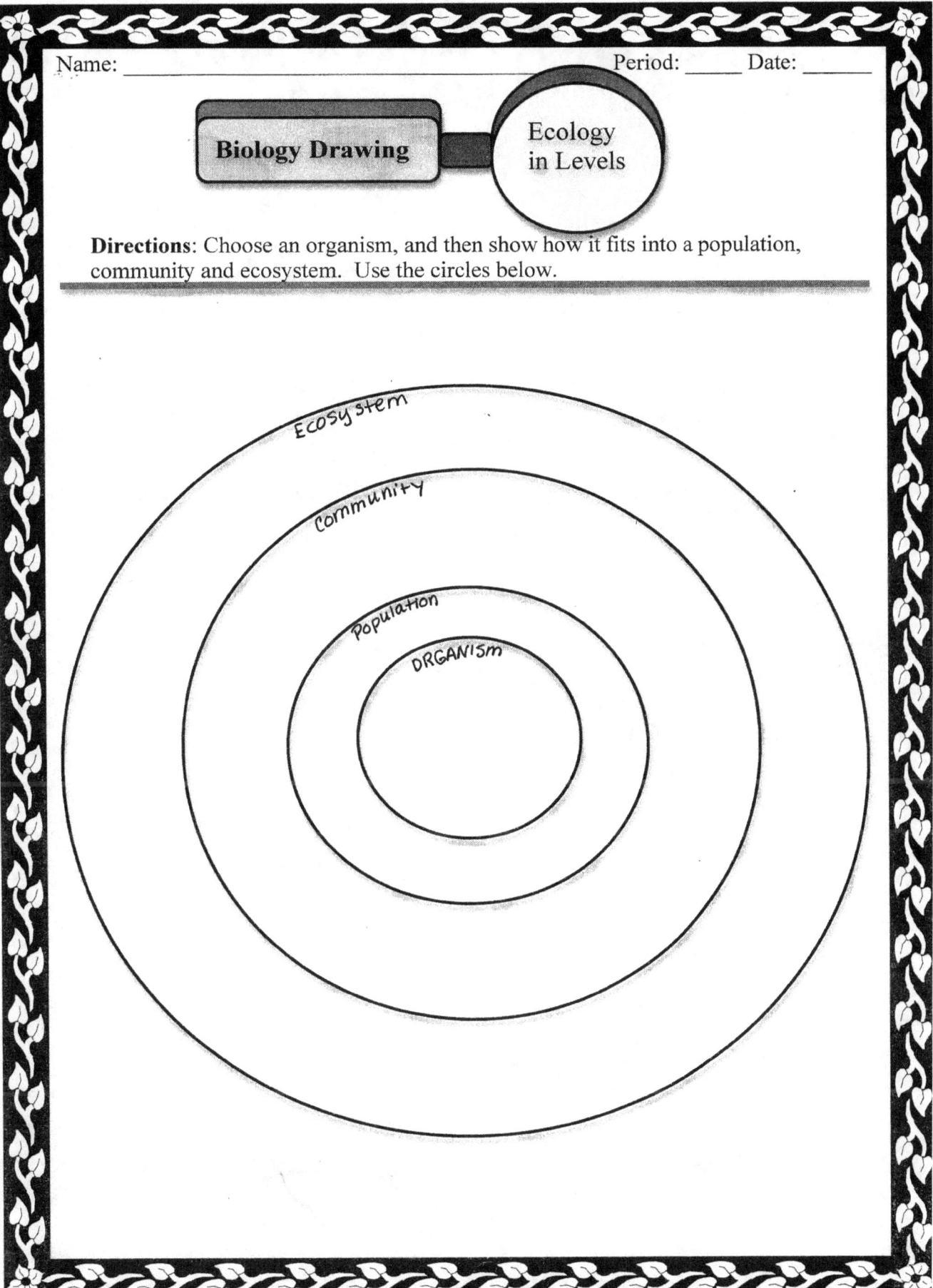

Name:_____ Period:_____ Date:_____

Biology Drawing — Species Interactions

Directions: Draw a "Wanted" poster that depicts any symbiotic relationship of your choice, such as parasitism, mutualism, commensalism, etc.

Name: _____   Period: _____ Date: _____

Biology Drawing   Food Web

Directions: Draw a food web containing the animal shown below. Be sure to show arrows going from one organism to the other.

Name: _____ Period: _____ Date: _____

Biology Drawing — Bio-Magnification

Directions: As we go up the food chain, we see that some harmful substances such as DDT gets more concentrated and thus is more magnified within the tissue. If we eat the animals at the top of the food chain, we will get a large concentration of these harmful substances along with problems associated with that intake. Draw an example of biomagnifications, denoting which substance you are depicting. Use the symbol in your drawing.

Name: _____ Period: _____ Date: _____

Biology Drawing — Biogeo-Chemical cycles

Directions: Combine the 3 types of cycles (water, carbon and nitrogen) in one picture. Label, and be sure to write the names of important molecules like carbon dioxide, water, etc. Be sure to include the symbol.

Name: _____ Period: _____ Date: _____

Biology Drawing — Greenhouse Effect

Directions: Draw the greenhouse effect using the sun and ground below. Be sure to show the greenhouse gases, sun's rays, etc.

75

Name: _____ Period: _____ Date: _____

Biology Drawing — Reduce, Reuse, and Recycle

Directions: Make an advertisement to encourage people to reduce, reuse and recycle. You can choose just one or all three.

Name: _____ Period: _____ Date: _____

Biology Drawing — Human Body Anatomy

Directions: In the outline below, draw as many major human body organs as you can in the correct position and proportion. Be sure to label the organs.

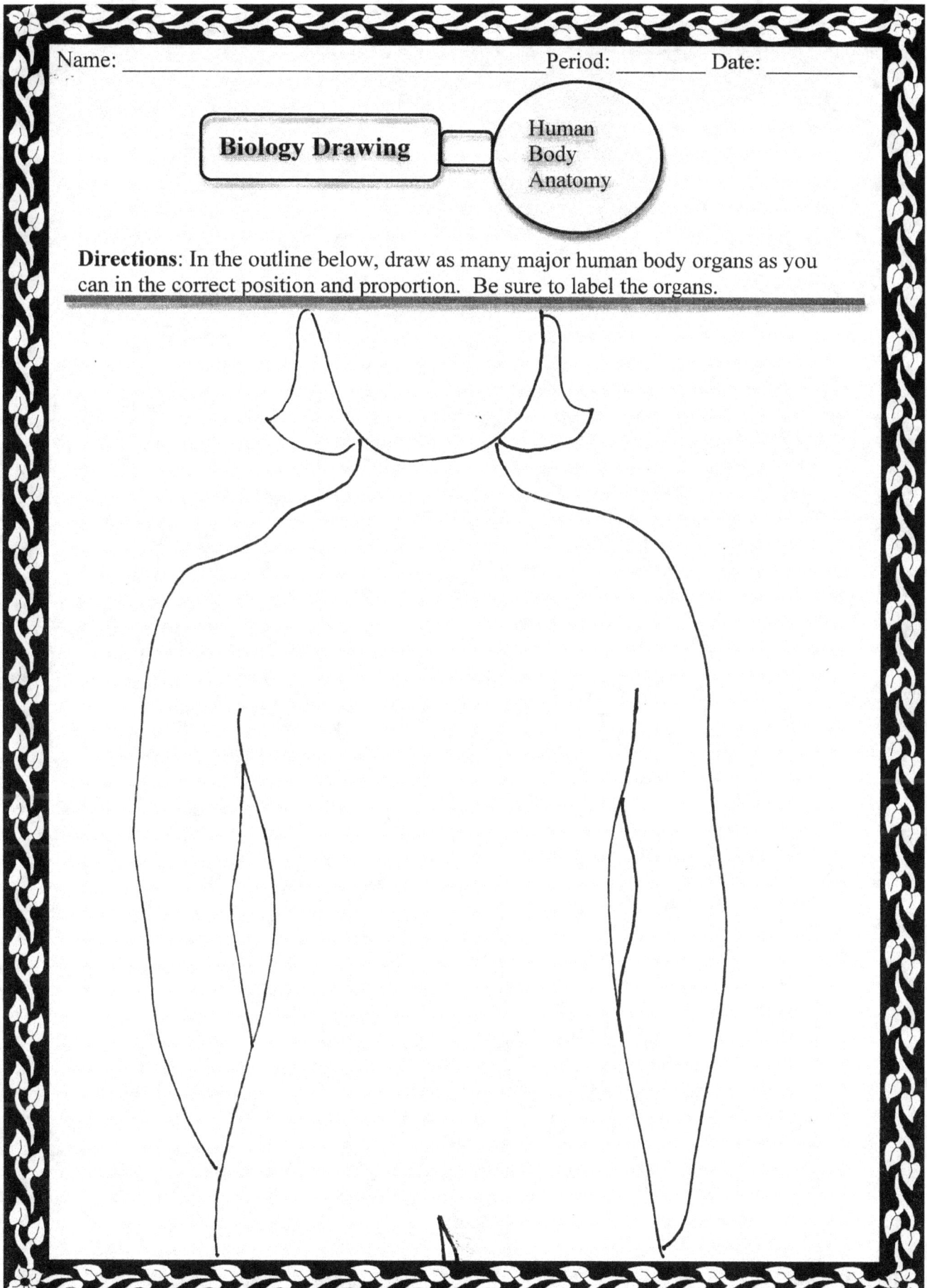

79

Biology Drawing — Nervous System

Directions: In the outline below, draw the components of the nervous system. Be sure to label the organs.

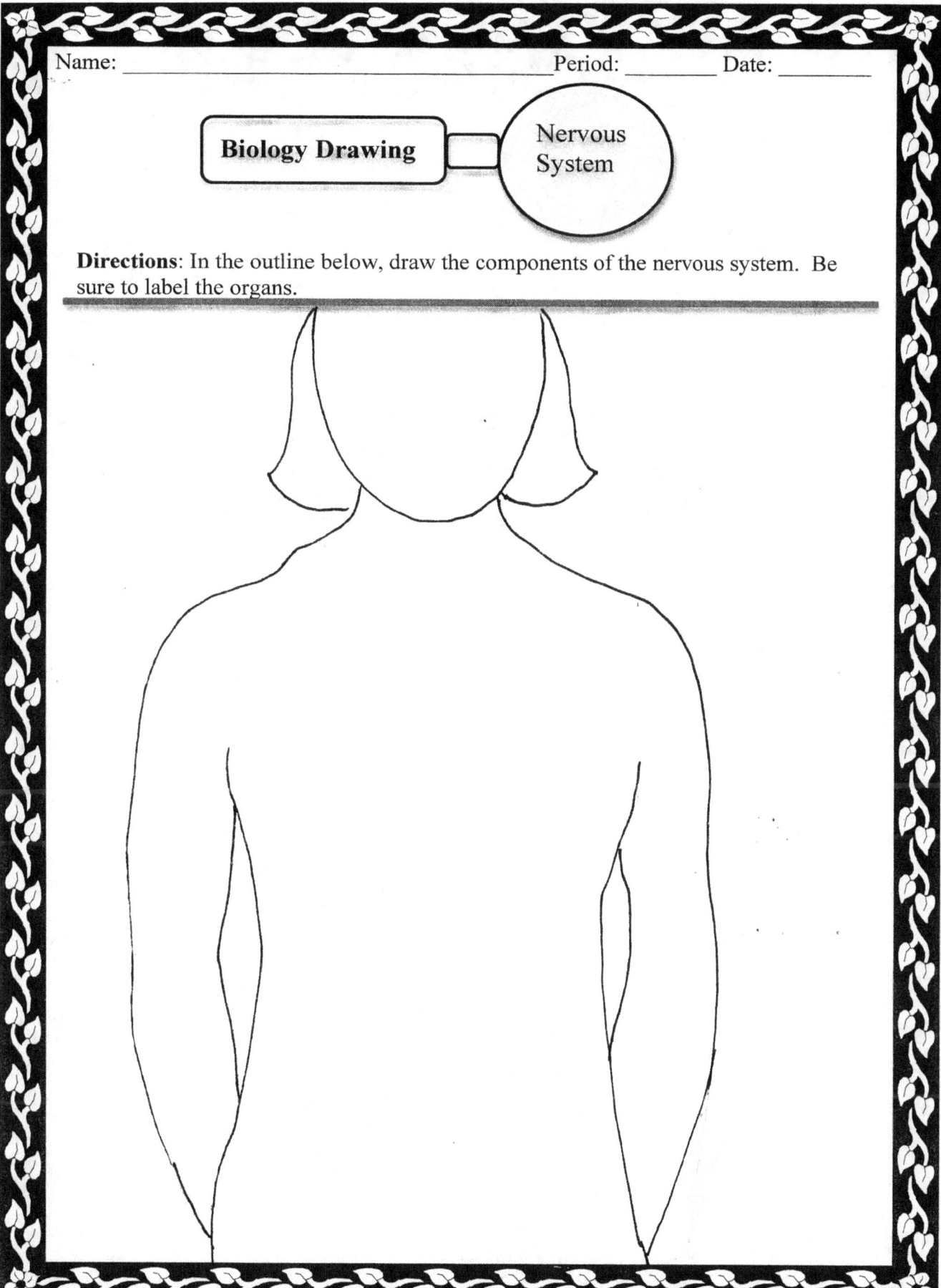

Name: _____ Period: _____ Date: _____

Biology Drawing — Endocrine System

Directions: In the outline below, draw the components of the endocrine system. Be sure to label the organs.

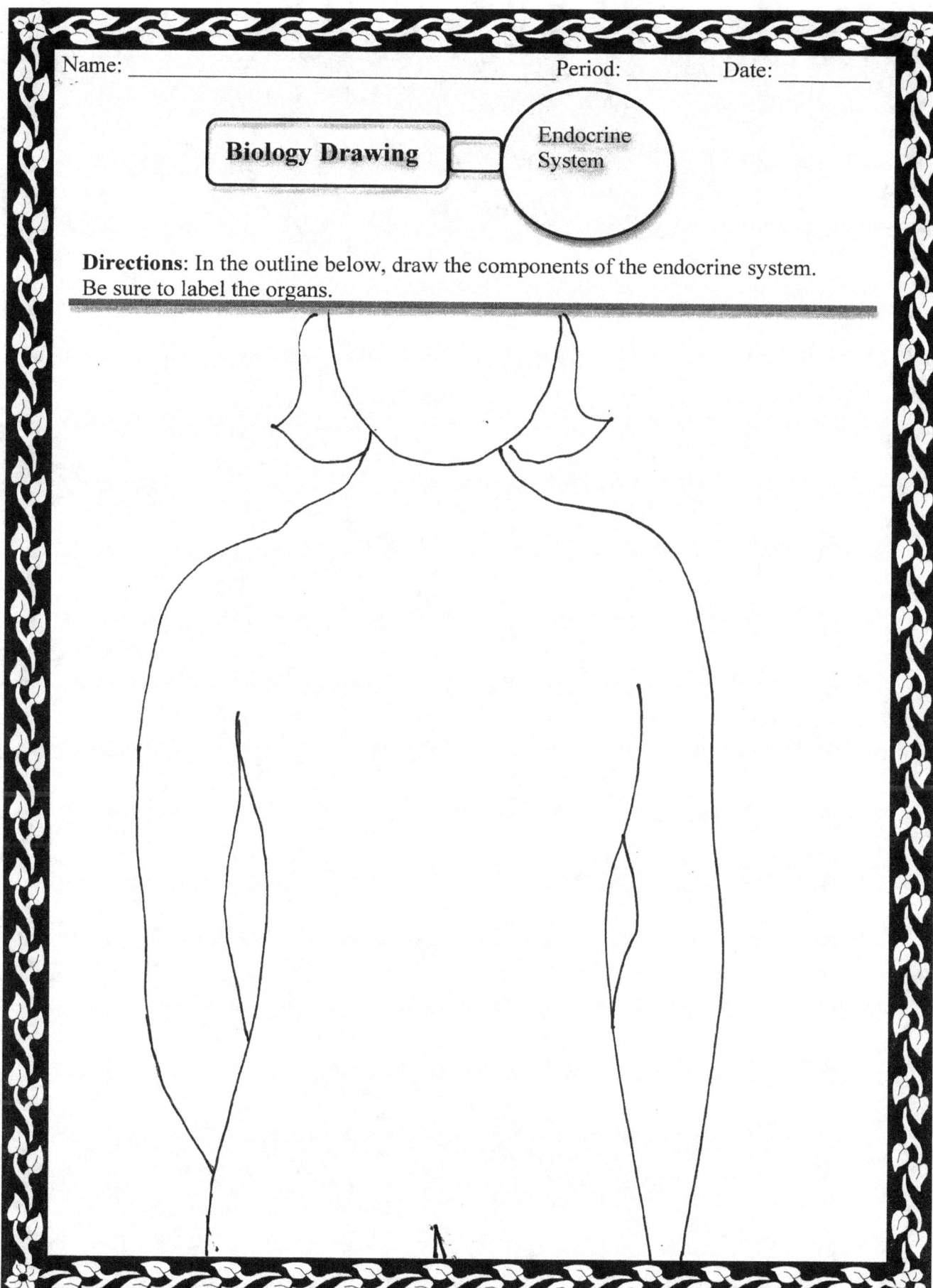

83

Name: _____ Period: _____ Date: _____

Biology Drawing — Immune System

Directions: In the outline below, draw the components of the endocrine system, including glands, organs, etc. Be sure to label the organs.

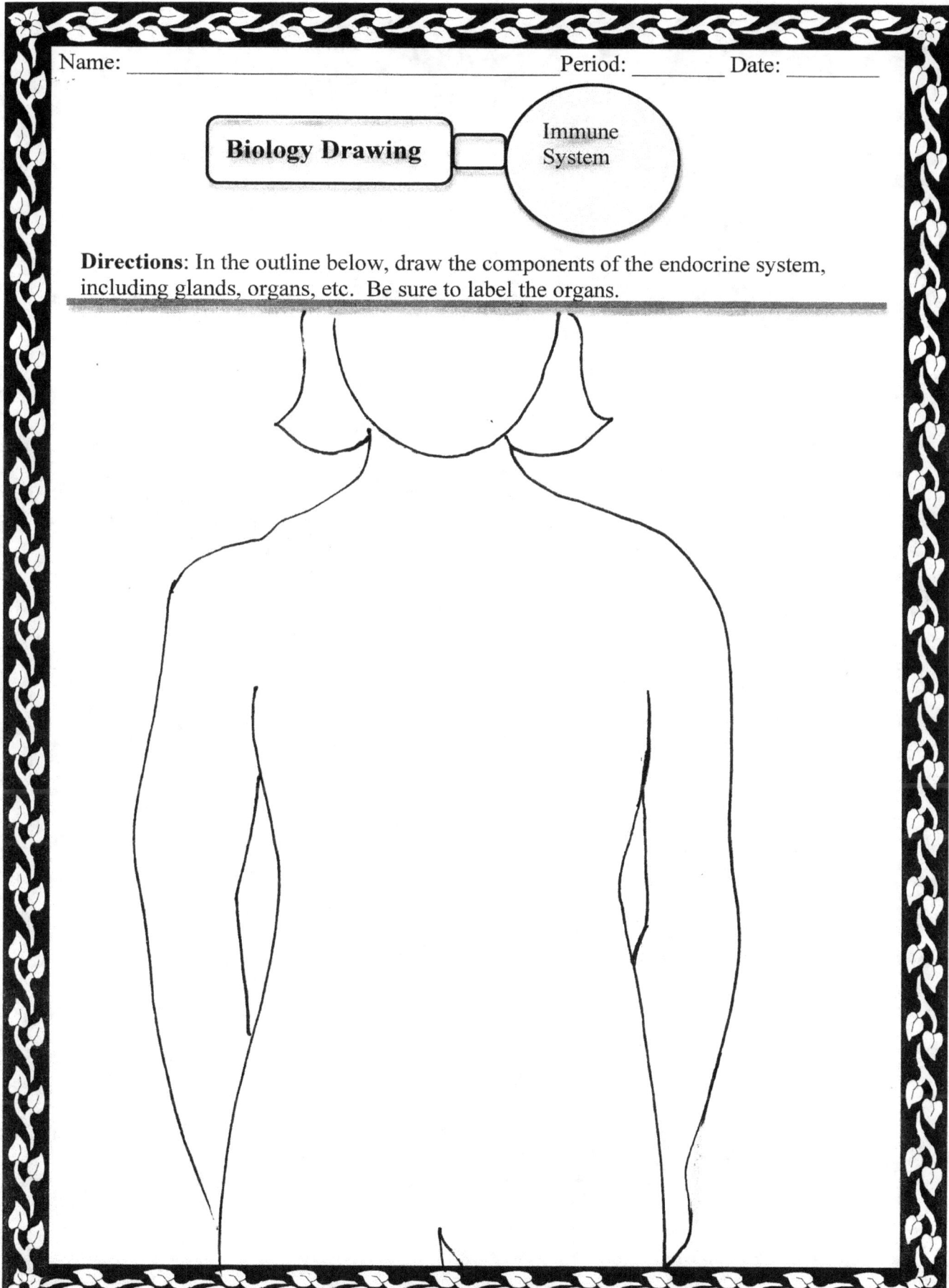

85

Name: _____ Period: _____ Date: _____

Biology Drawing — Integumentary System

Directions: Draw a section of the human skin, showing the different layers and important structures. Use the line below in your drawing.

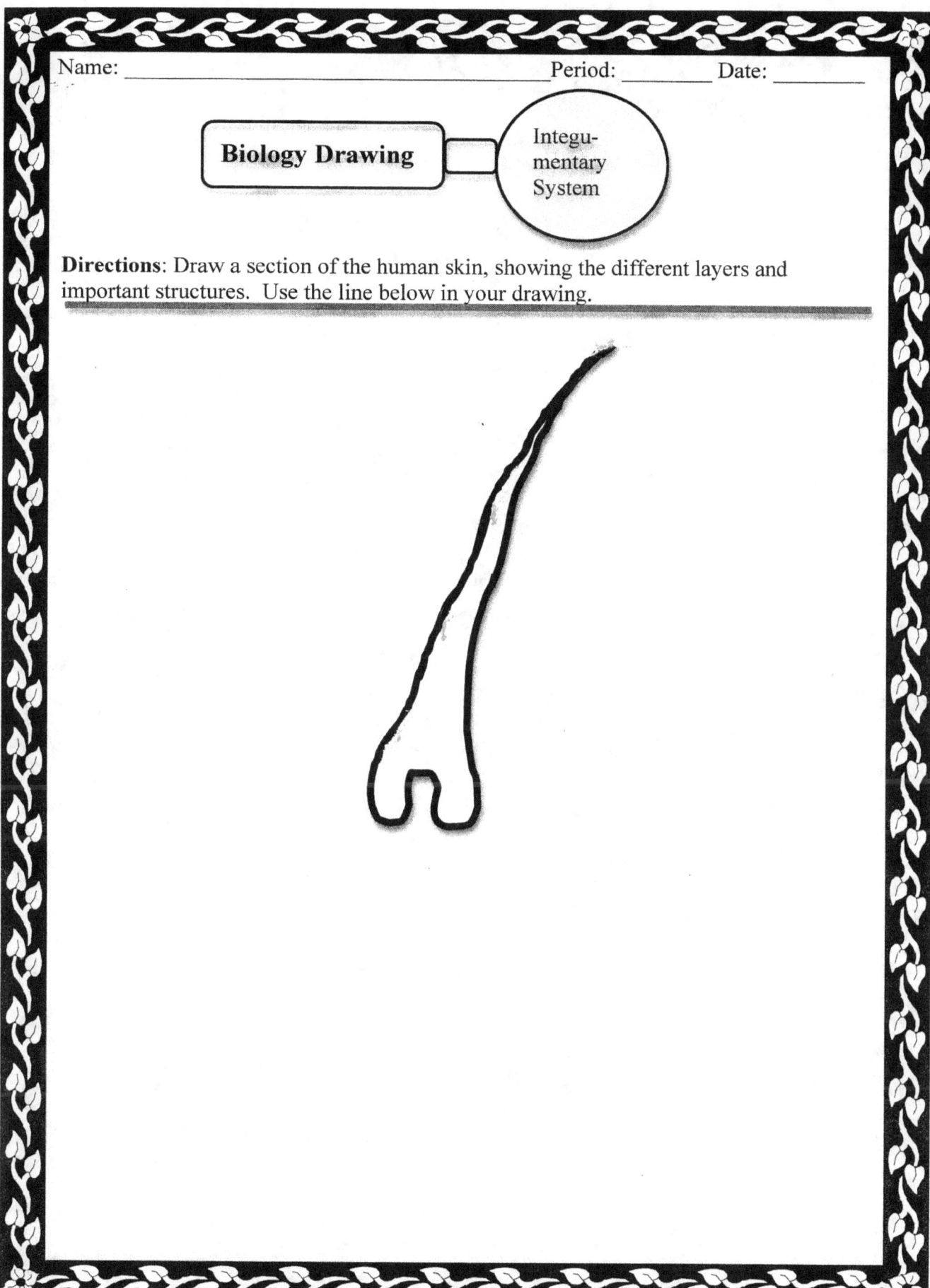

Name: _____ Period: _____ Date: _____

Biology Drawing — Digestive System

Directions: In the outline below, draw all of the digestive organs, label and color. Then, if directed to, show the three types of food molecules being digested along the way. Use circles for carbohydrate monomers, squares for proteins (amino acids) and triangles for fats.

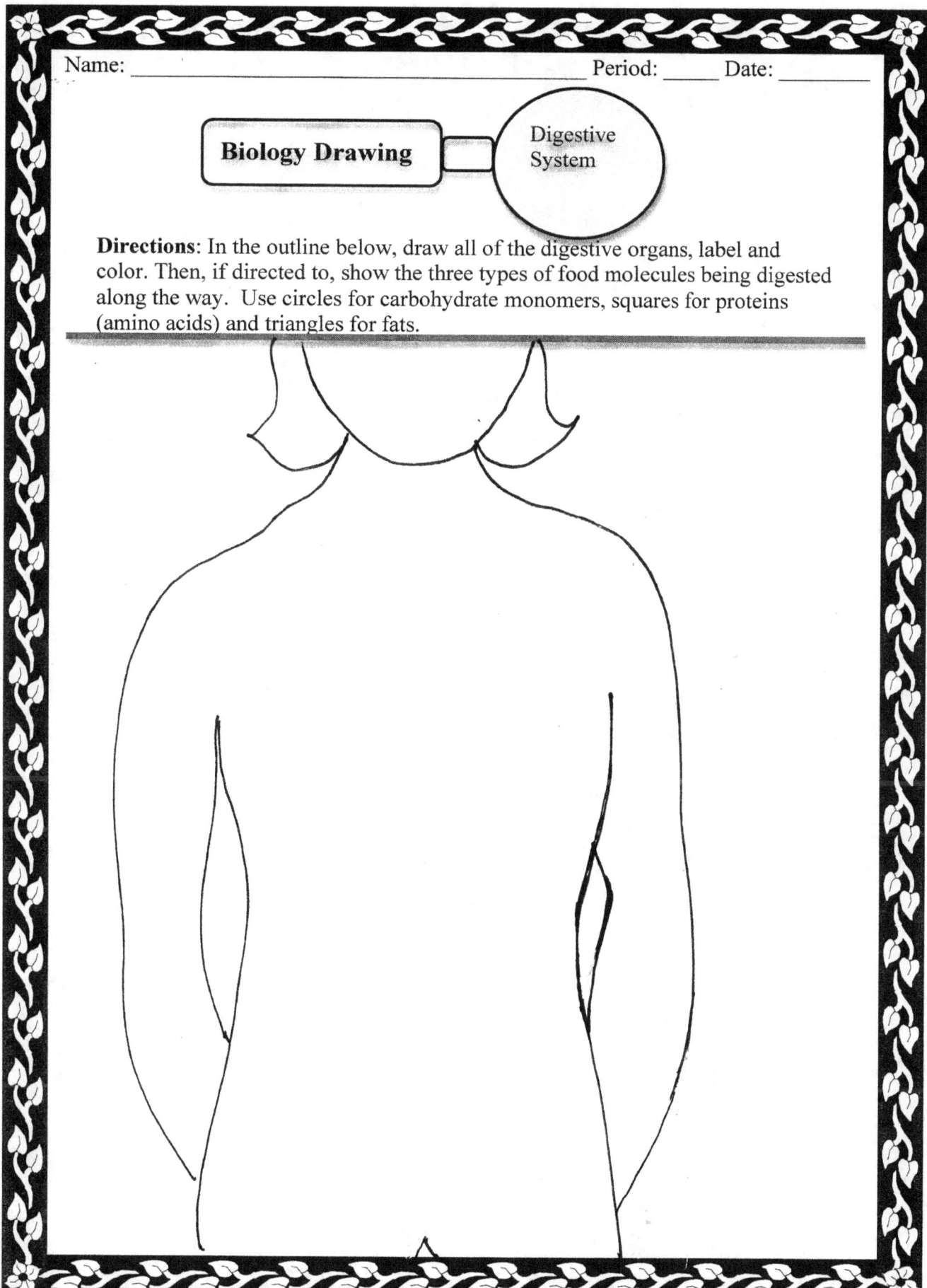

89

Name: _____ Period: _____ Date: _____

Biology Drawing — Choose a Body Organ

Directions: Draw a body organ of your choice showing as much detail as required by your teacher. Be sure to label and color. On the other side, explain where the organ is found in the body, what it's job is, and whether it is essential to body function. Use the symbol below in the drawing.

Name: _____ Period: _____ Date: _____

Biology Drawing — Anatomy of the Frog

Directions: Draw a diagram of the frog, showing it's internal anatomy. Be sure to label and color.

Student/Teacher Samples

Name: _____ Period: _____ Date: _____

Biology Drawing — Art Forms in Nature

Directions: Find a natural object, such as a leaf or insect. Draw it and color it in an artistic way to showcase art in nature as done by many scientists in the past

Name: _____ Period: _____ Date: _____

Biology Drawing — The 6 Kingdoms of Life

Directions: Divide the space below into 6 sections. Label each section with one of the 6 kingdoms. Draw a representative of each kingdom, name it, and write down the characteristics of each kingdom in the space. You may add cut-out drawings to make it look like a collage if your teacher so desires.

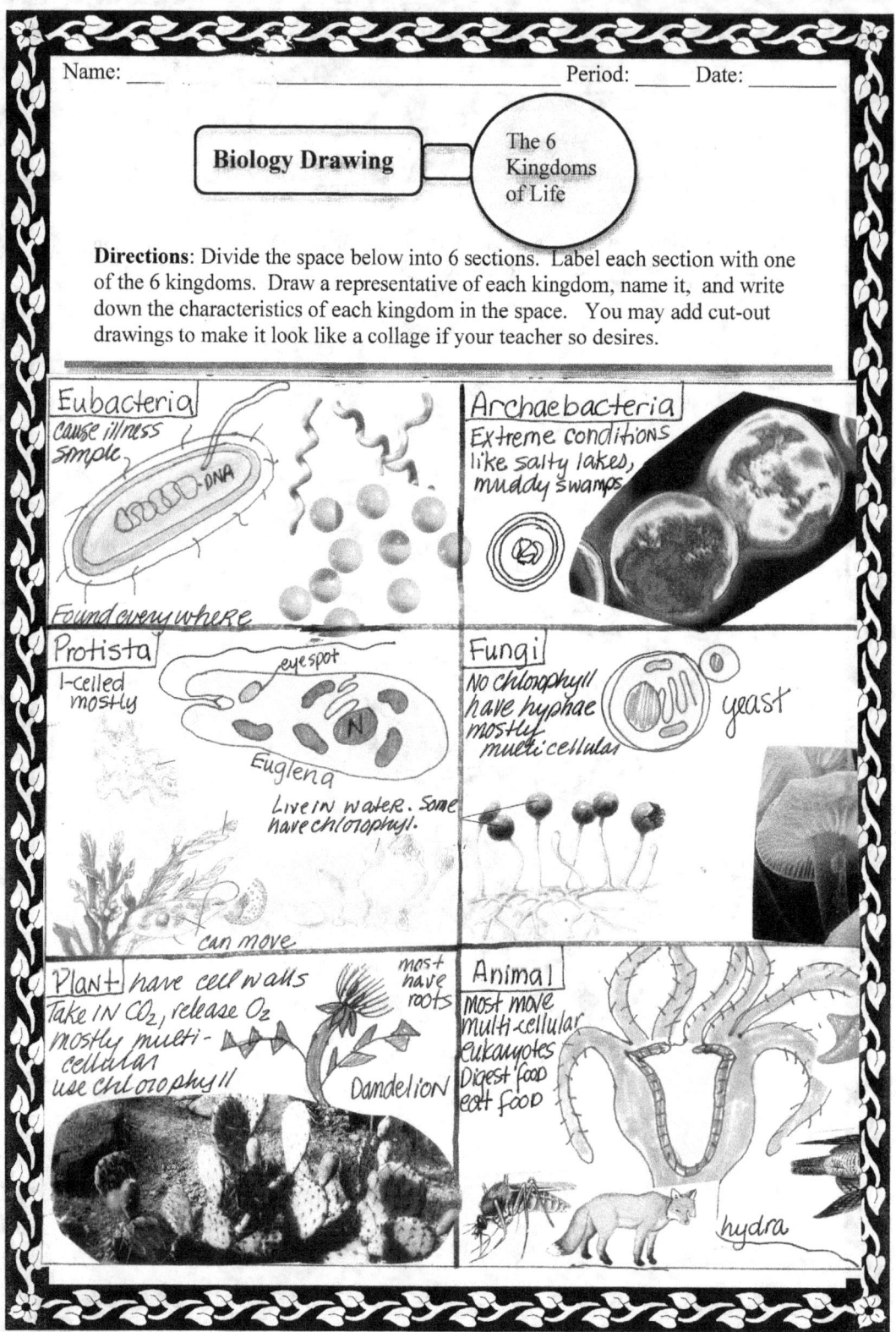

96

Name: _____   Period: _____ Date: _____

Biology Drawing — Enzyme Action

Directions: Draw an enzyme showing the active site and then use the molecules shown to fit into the active site. The enzyme can be used to bring the two molecules together or to split one molecule into two. Label.

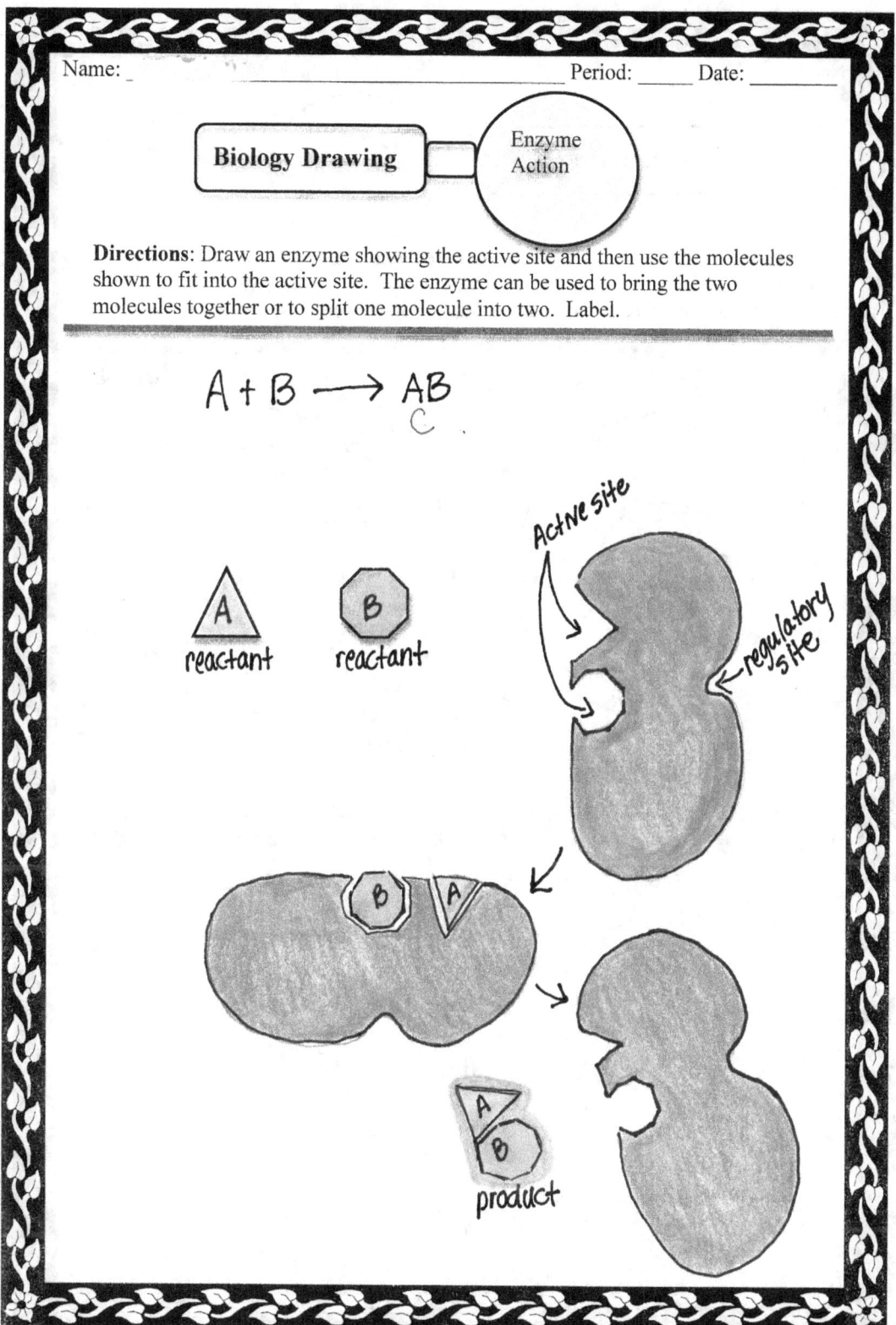

Name: _____ Period: _____ Date: _____

Biology Drawing — Elements In Organisms

Directions: Divide the space below into 4 sections where you will draw a carbon, hydrogen, nitrogen and oxygen atoms which constitute 95% of your body weight. Be sure to show the nucleus, orbitals, electrons, protons & neutrons. Label and color. Use the circle below in your picture.

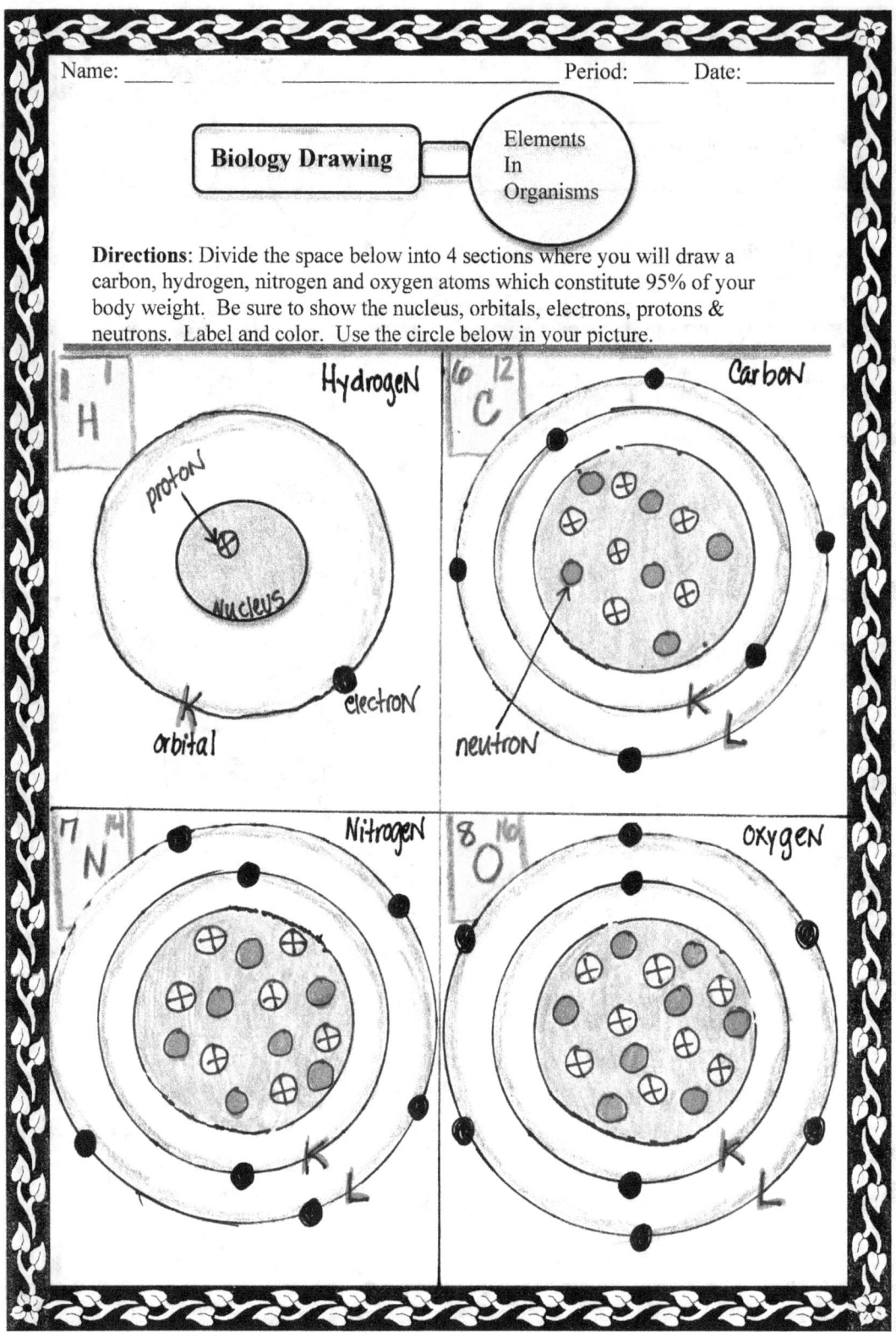

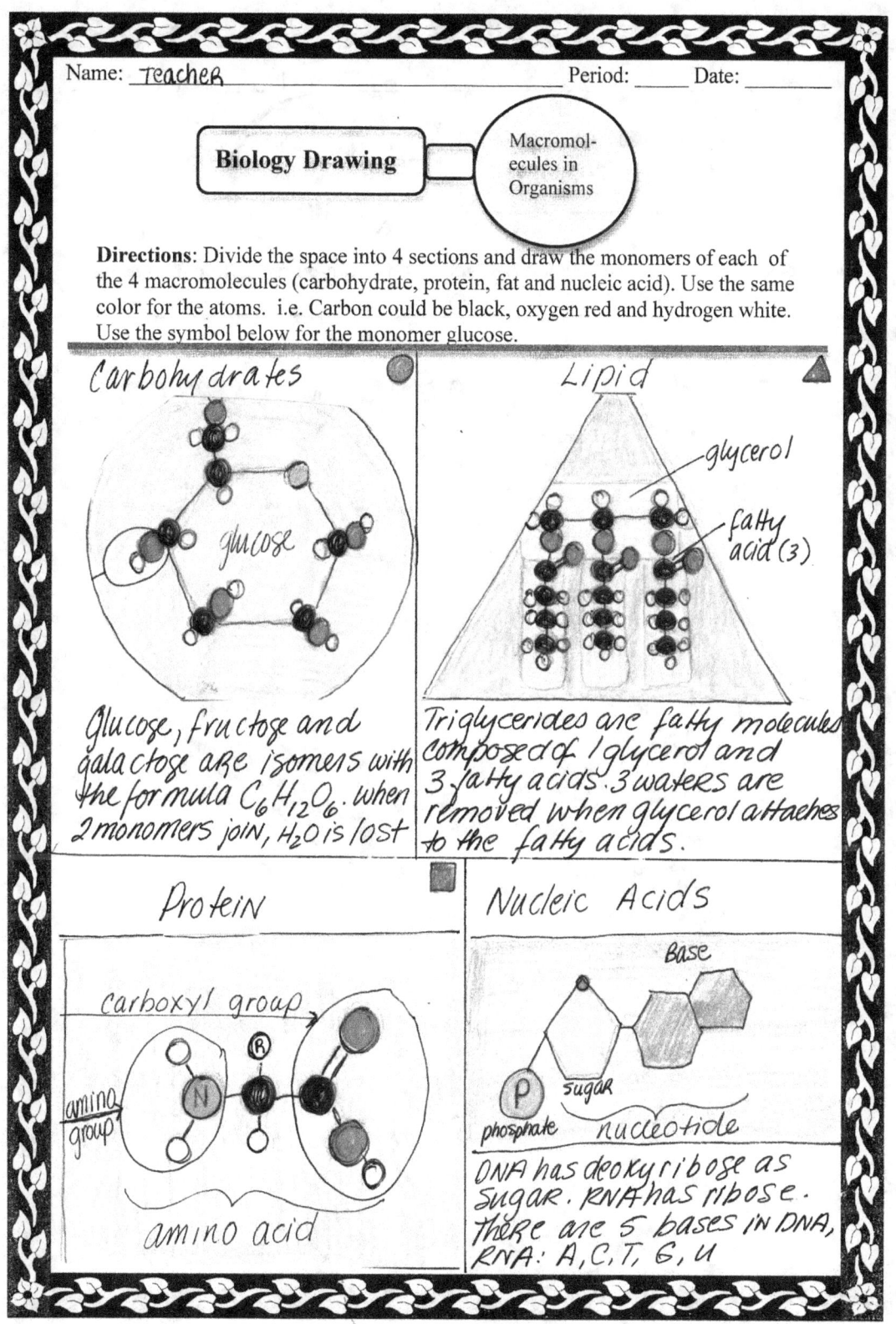

Name:_____ Period:____ Date:_____

Biology Drawing — Heirarchy of Life

Directions: Draw a way to show the heirarchy of life from **organelle** to **cell** to **tissue** to **organ** to **organ system** and finally, to **body**.

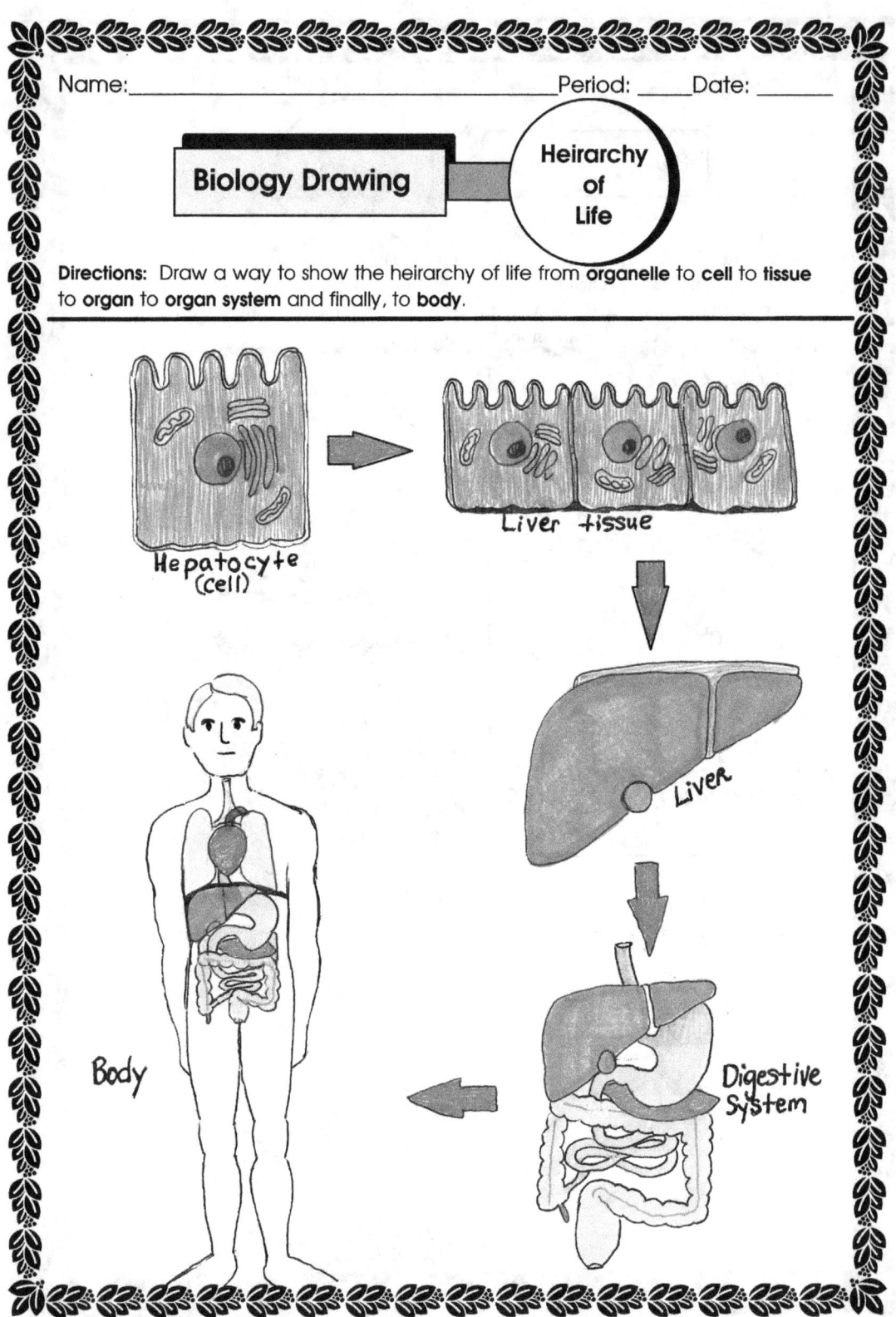

Hepatocyte (cell) → Liver tissue → Liver → Digestive System → Body

Name: _Teacher_   Period: ____   Date: ____

Biology Drawing — Cell Analogy

Directions: The cell is often compared to towns, schools, factories and other common systems. Pick one of these or something else to compare the cell to. Divide the space in half, showing the cell in one half, and the analogy in the other, using the same colors for comparison. i.e. the mitochondria and the town's powerplant would both be colored brown. Be sure to label. Use the symbol below in your drawings.

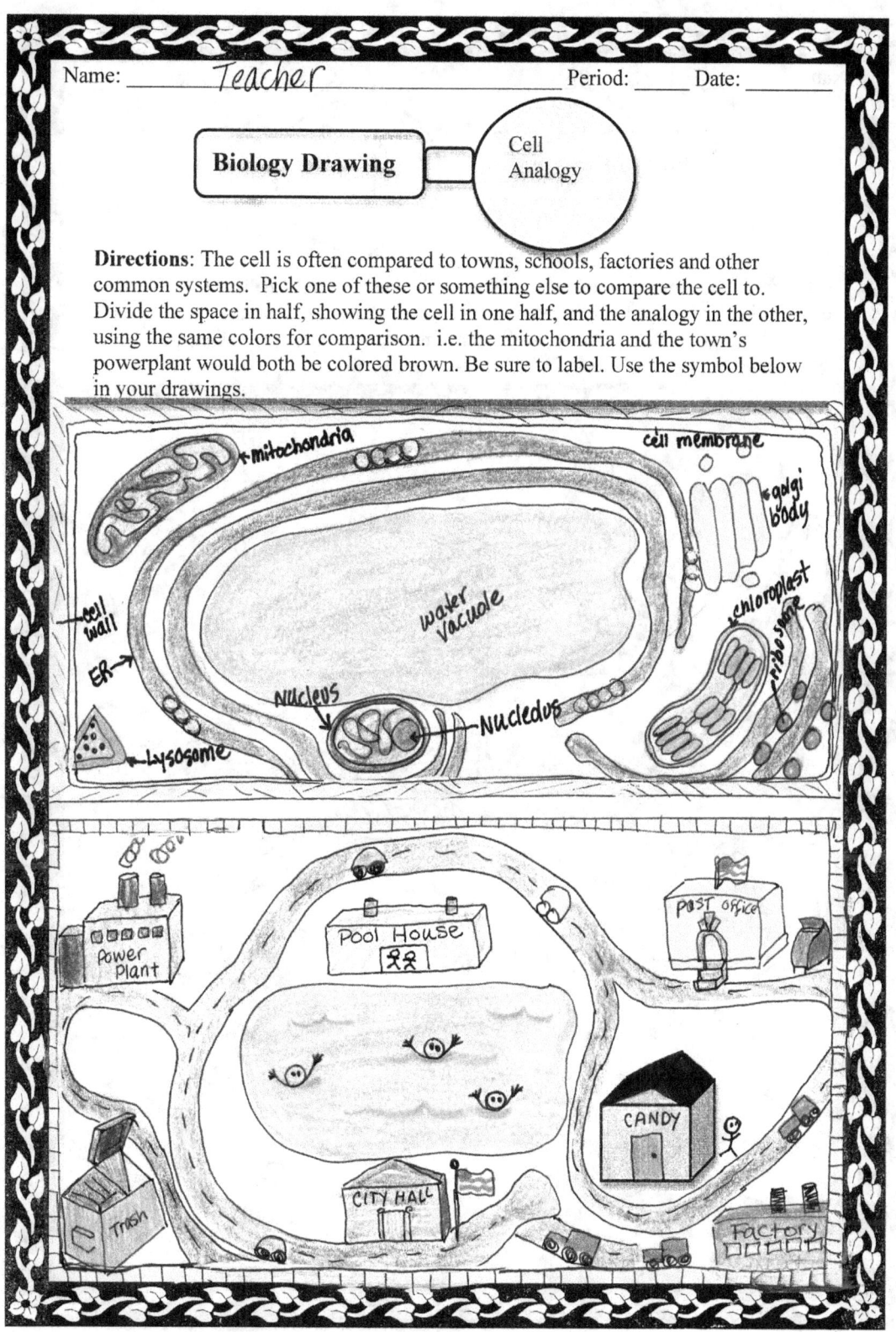

Name: _teacher_ Period: ____ Date: ____

Biology Drawing — Cell Types

Directions: Divide the space into 4 sections. Label the sections: animal cell, plant cell, protist cell, and bacterial cell (prokaryotic). The first three cells are eukaryotic. Draw a general representative of each type of cell, and then color code the drawings so that the same organelles have the same colors throughout. Use the symbol below in your drawing.

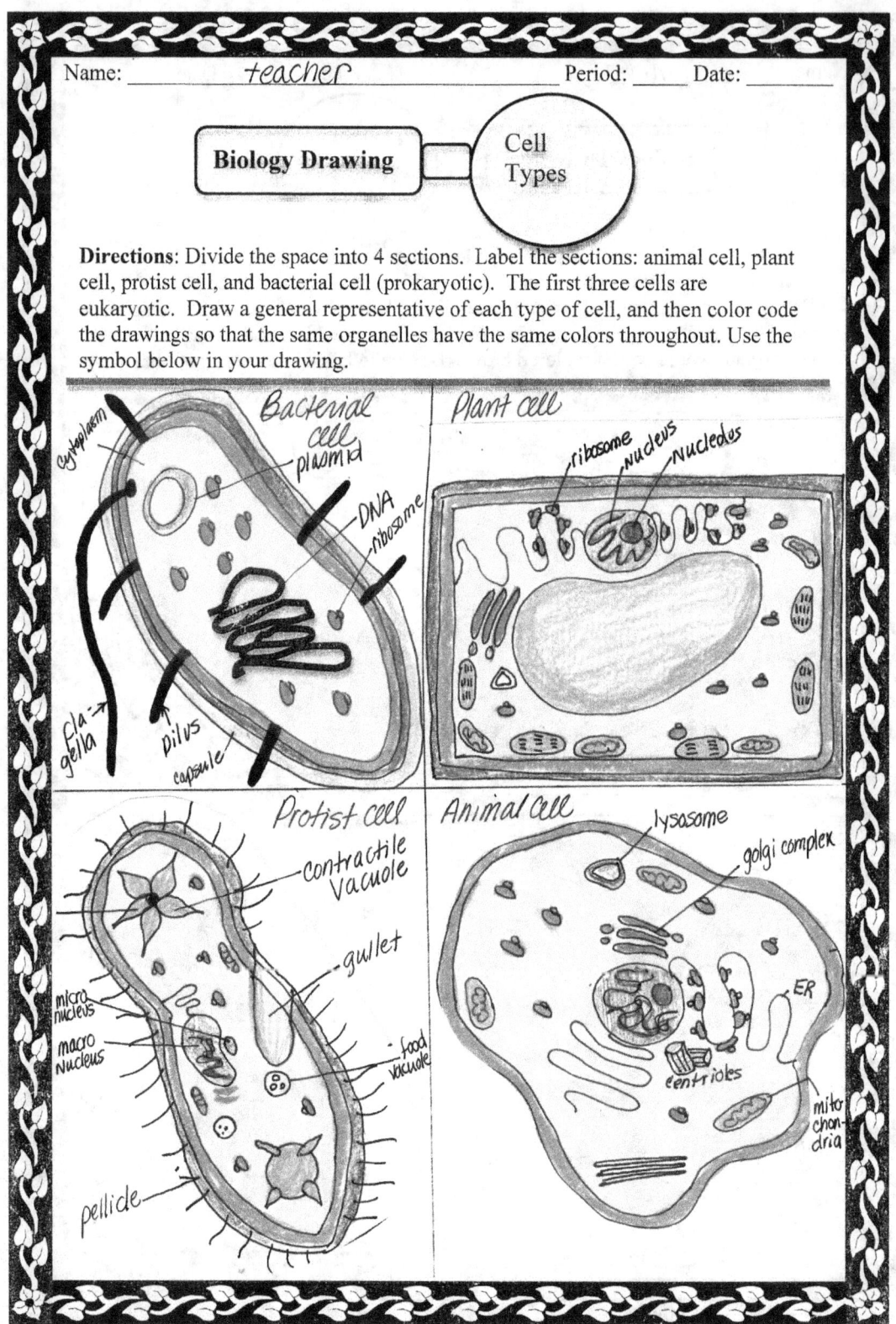

Name: Student   Period: ____ Date: _____

Biology Drawing — Homeostasis

Directions: Draw a representation of a negative feedback loop, showing a real-life example. Don't forget to label all the steps in the circular loop below.

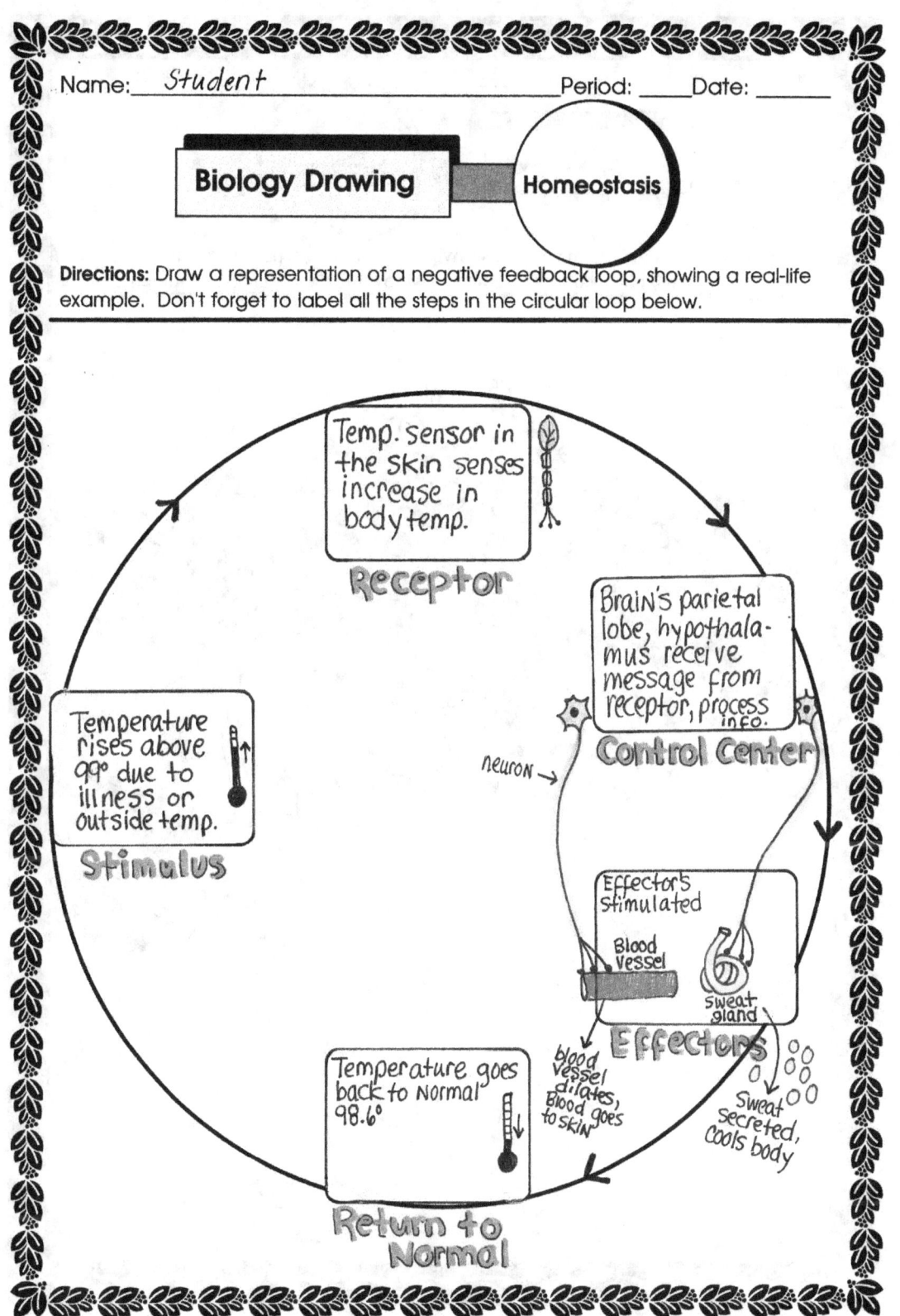

- **Receptor:** Temp. sensor in the skin senses increase in body temp.
- **Control Center:** Brain's parietal lobe, hypothalamus receive message from receptor, process info. (neuron)
- **Effectors:** Effectors stimulated — Blood vessel, sweat gland. Blood vessel dilates, blood goes to skin; sweat secreted, cools body.
- **Return to Normal:** Temperature goes back to normal 98.6°
- **Stimulus:** Temperature rises above 99° due to illness or outside temp.

Name: Teacher   Period: ____ Date: ____

Biology Drawing — The Cell Membrane

Directions: Draw a cell membrane showing the different types of proteins embedded in the phospholipids. Then, show movement through the membrane.

Name: student   Period: __ Date: __

Biology Drawing — The Cell Membrane

Directions: Draw a cell membrane showing the different types of proteins embedded in the phospholipids. Then, show movement through the membrane.

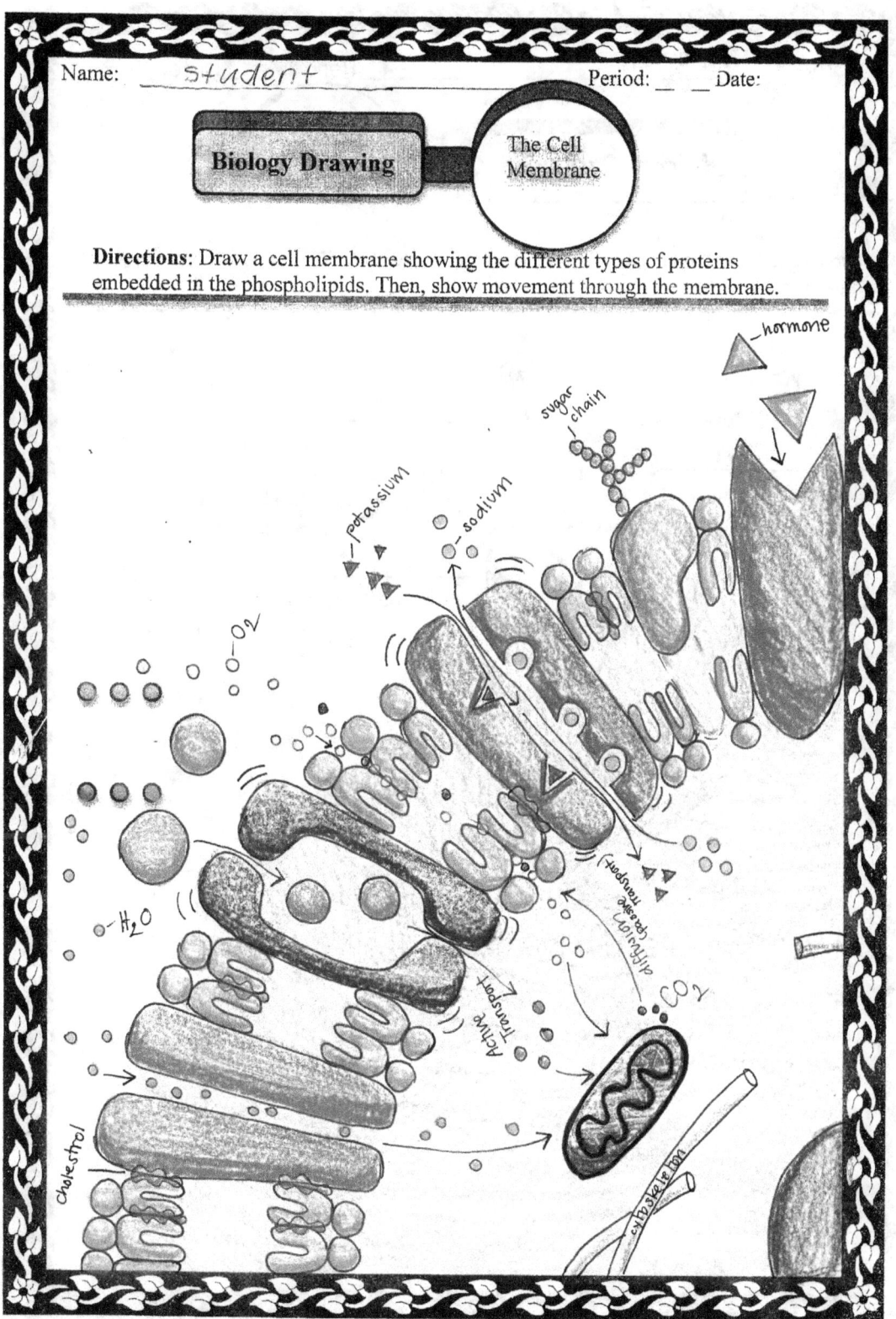

Name: __Student__ Period: ____ Date: ____

Biology Drawing — ATP

Directions: Draw the structure of ATP and show how it stores energy within it's high energy bonds.

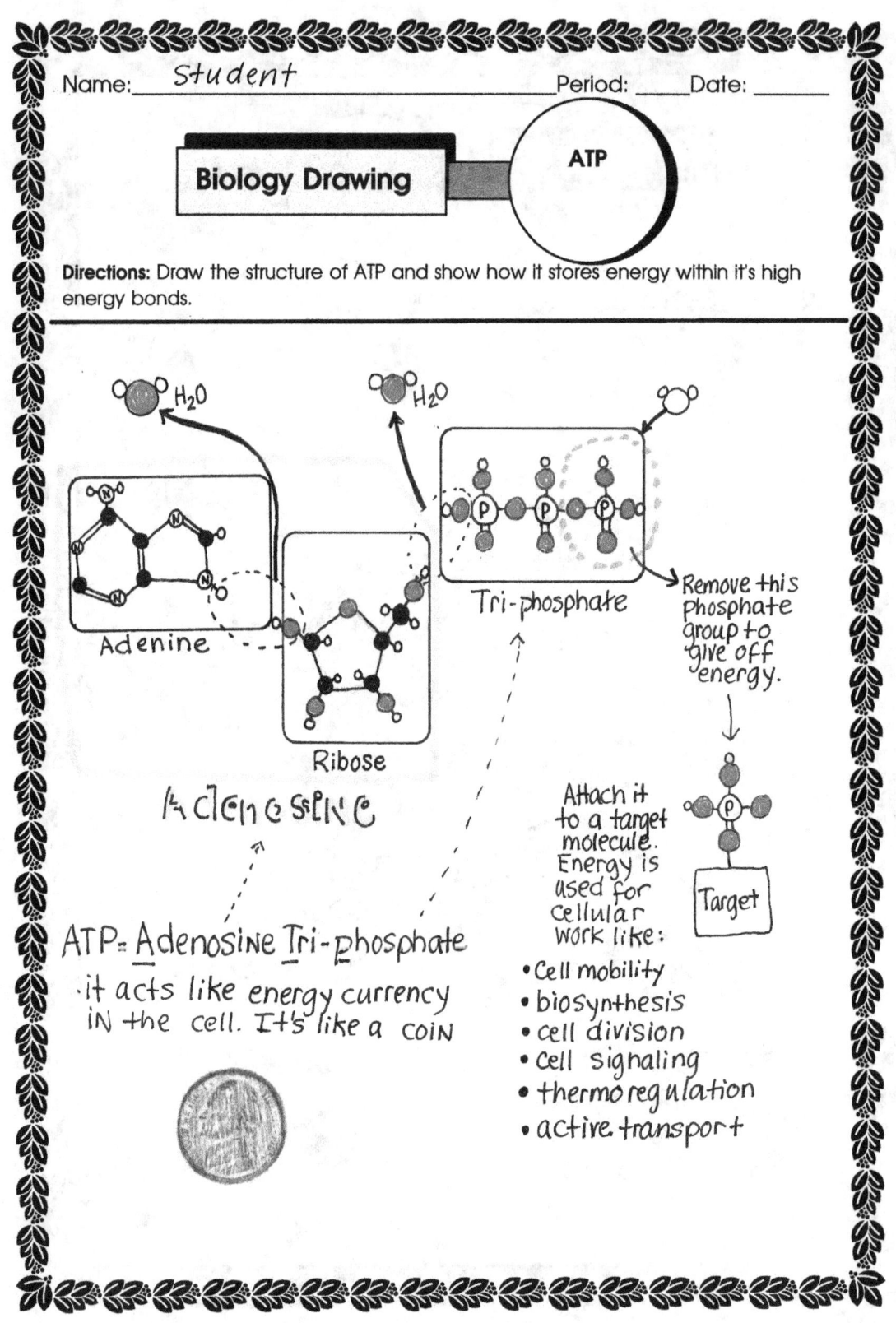

ADENOSINE

ATP = <u>A</u>denosine <u>T</u>ri-<u>p</u>hosphate

- it acts like energy currency in the cell. It's like a coin

- Cell mobility
- biosynthesis
- cell division
- cell signaling
- thermoregulation
- active transport

Name: teacher   Period: ___ Date: ___

Biology Drawing — Life Cycle

Directions: Using the circle below, draw the components of the life cycle, showing how animals and plants keep each other alive with the exchange of gases and other molecules.

Photosynthesis: ☼ + $6CO_2 + 6H_2O \rightarrow 6O_2 + C_6H_{12}O_6$

Cellular Respiration: $C_6H_{12}O_6 + 6O_2 \rightarrow 6CO_2 + 6H_2O + 38\ ATP$

Name: Teacher   Period: ___ Date: ___

Biology Drawing — Photo-synthesis

Directions: Draw a simplified view of photosynthesis, giving as much detail as your teacher requests. Be sure to label and color, using the sun below.

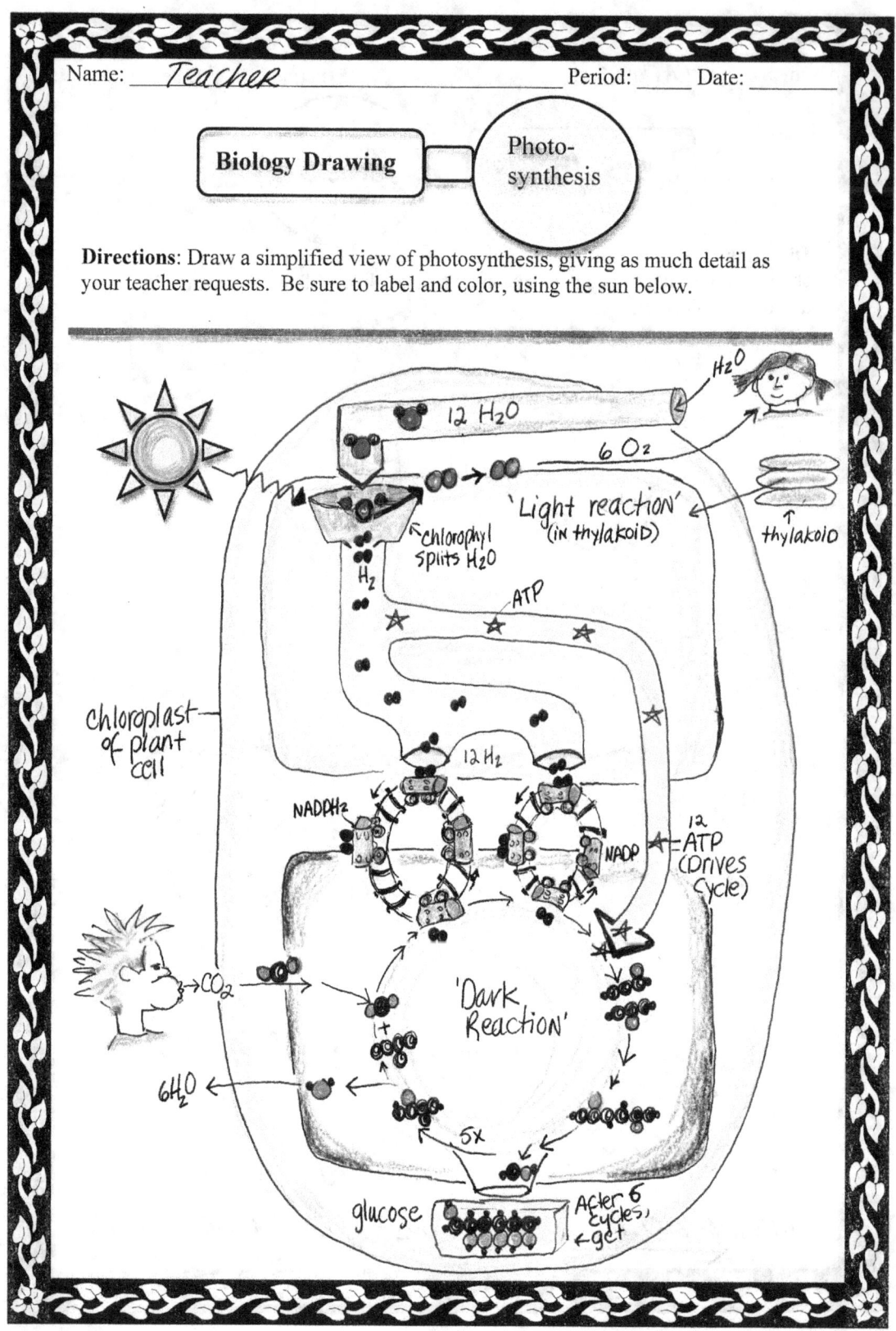

108

Biology Drawing — Cellular Respiration

Directions: Draw a simplified view of cellular respiration, giving as much detail as your teacher requests. Be sure to label and color, using the circle below for the Kreb's cycle

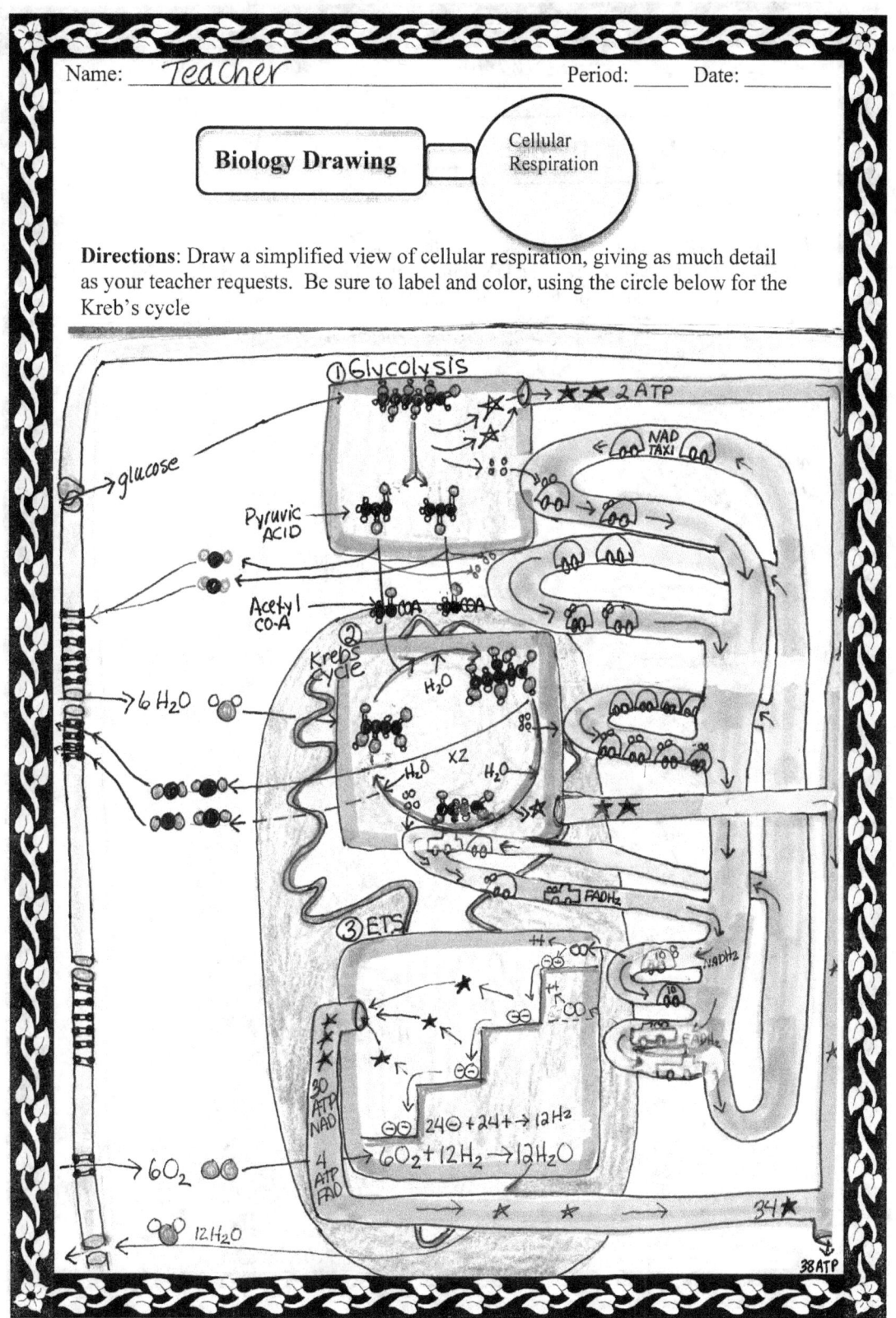

Name: _____Student_____ Period: _____ Date: _____

Biology Drawing — The Chromosome

Directions: Draw a chromosome below with as much detail as you can. If directed by your teacher, draw the DNA inside the chromosome and show how it wraps around the histone proteins. Label, color and use the symbol below.

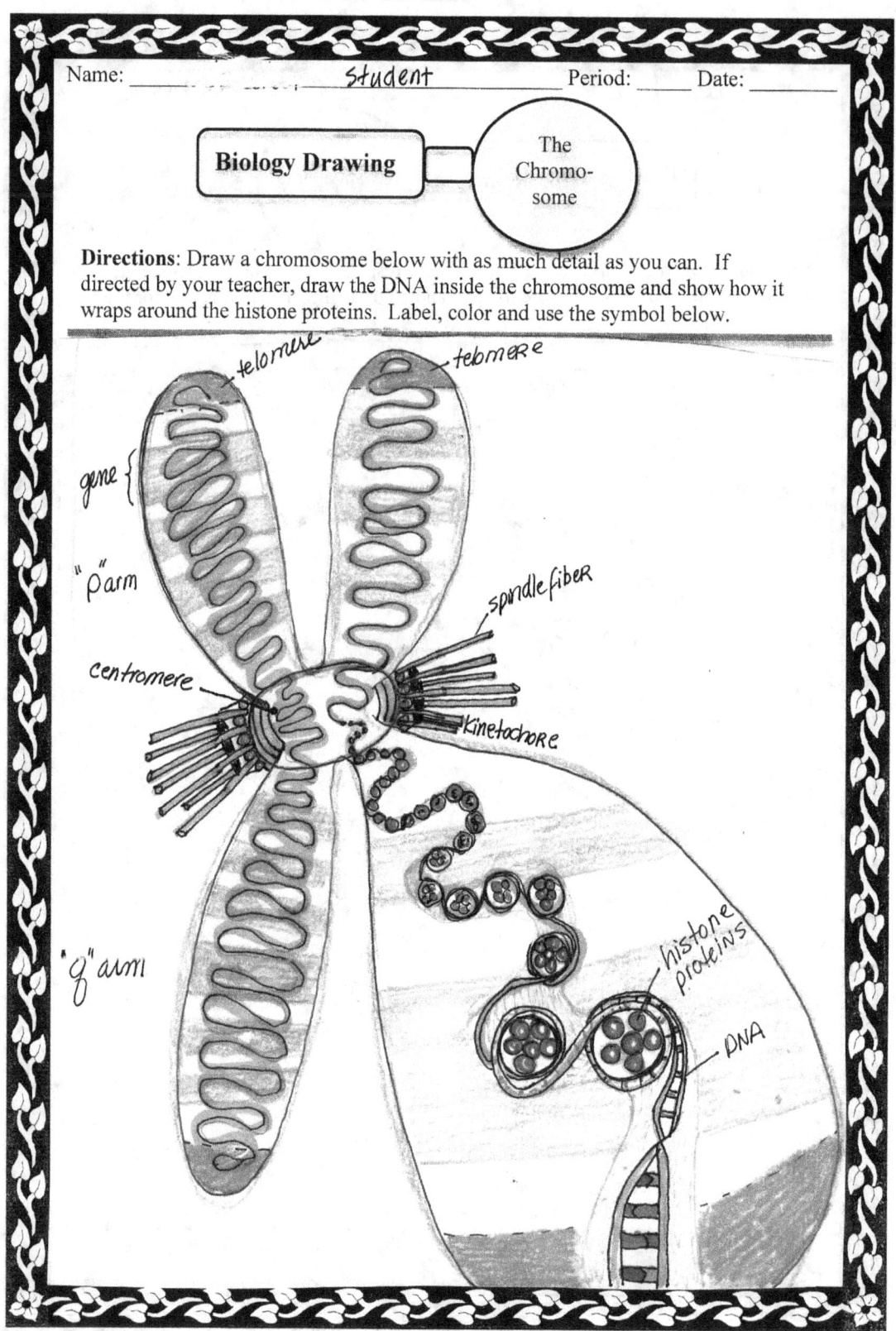

Name: **Teacher** Period: ___ Date: ___

Biology Drawing — The Cell Cycle

Directions: Use the circle below to draw the cell cycle. Use a ruler to divide it into the phases of the cycle. Label the phases and draw what the chromosome looks like in each phase and tell what is happening in each phase.

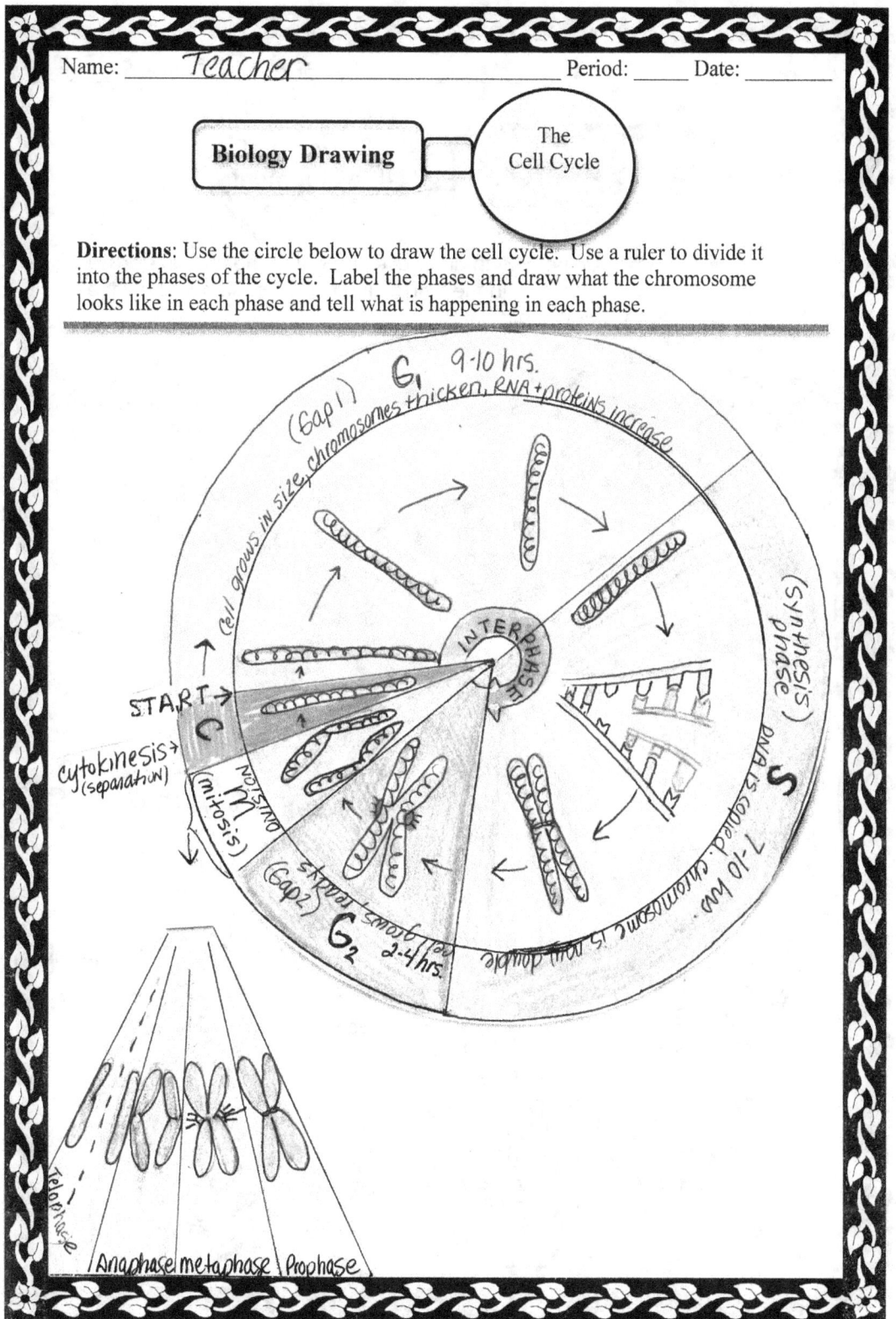

Name: Teacher Period: ____ Date: ____

Biology Drawing — Growth and Development Timeline

Directions: Draw a timeline of the growth and development of the human starting with the fertilized egg and ending with the birth of the baby. Be sure to show the morula, blastocyst, fetus, embryo, etc.

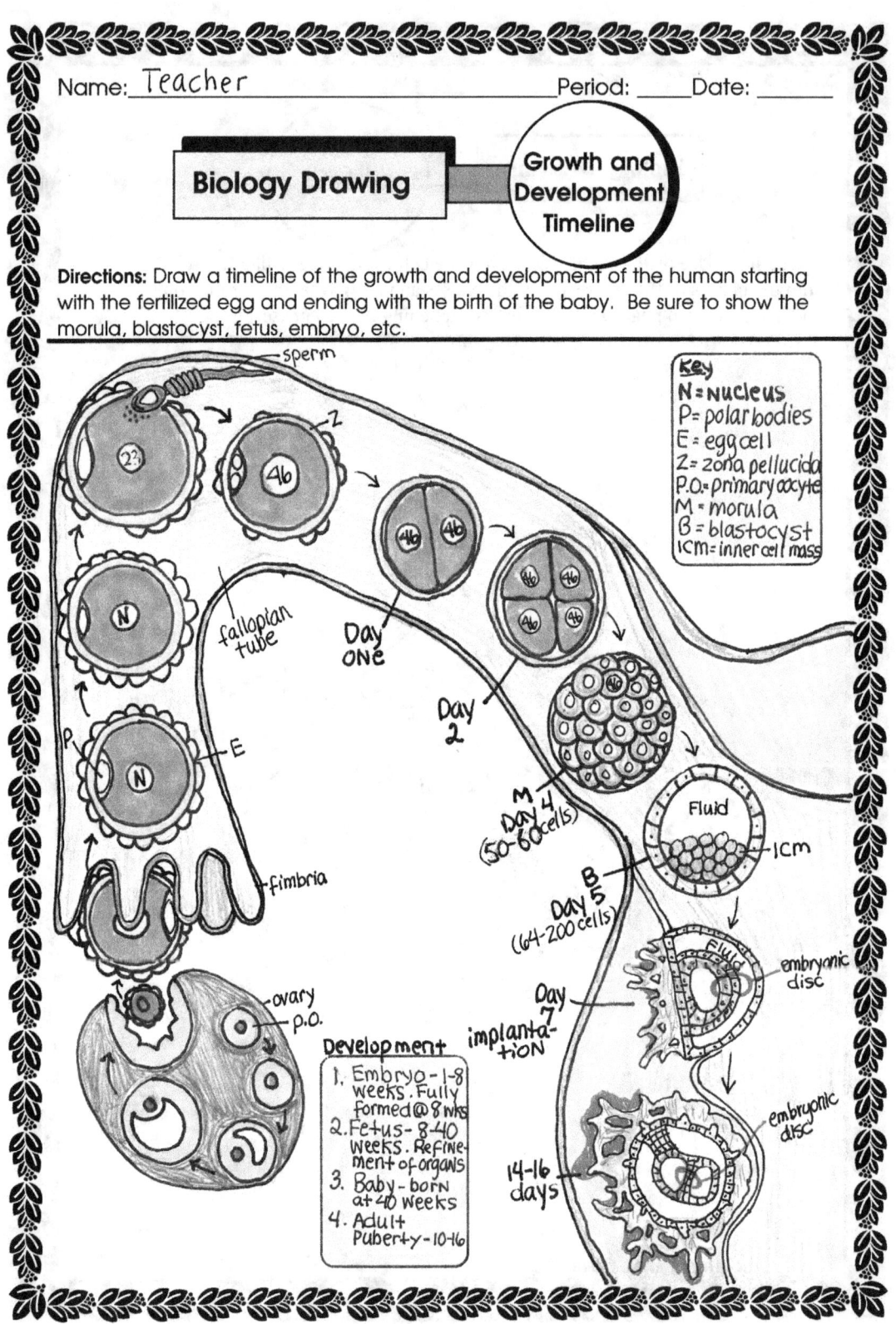

Name: _teacher_   Period: ____ Date: ____

Biology Drawing — Mitosis

Directions: Draw a cell undergoing mitosis, showing all of the steps and labeling each. Use at least 4 chromosomes in the drawing (2 homologous pairs). Use different colors for each pair. Show spindle fibers also. Use the symbol below.

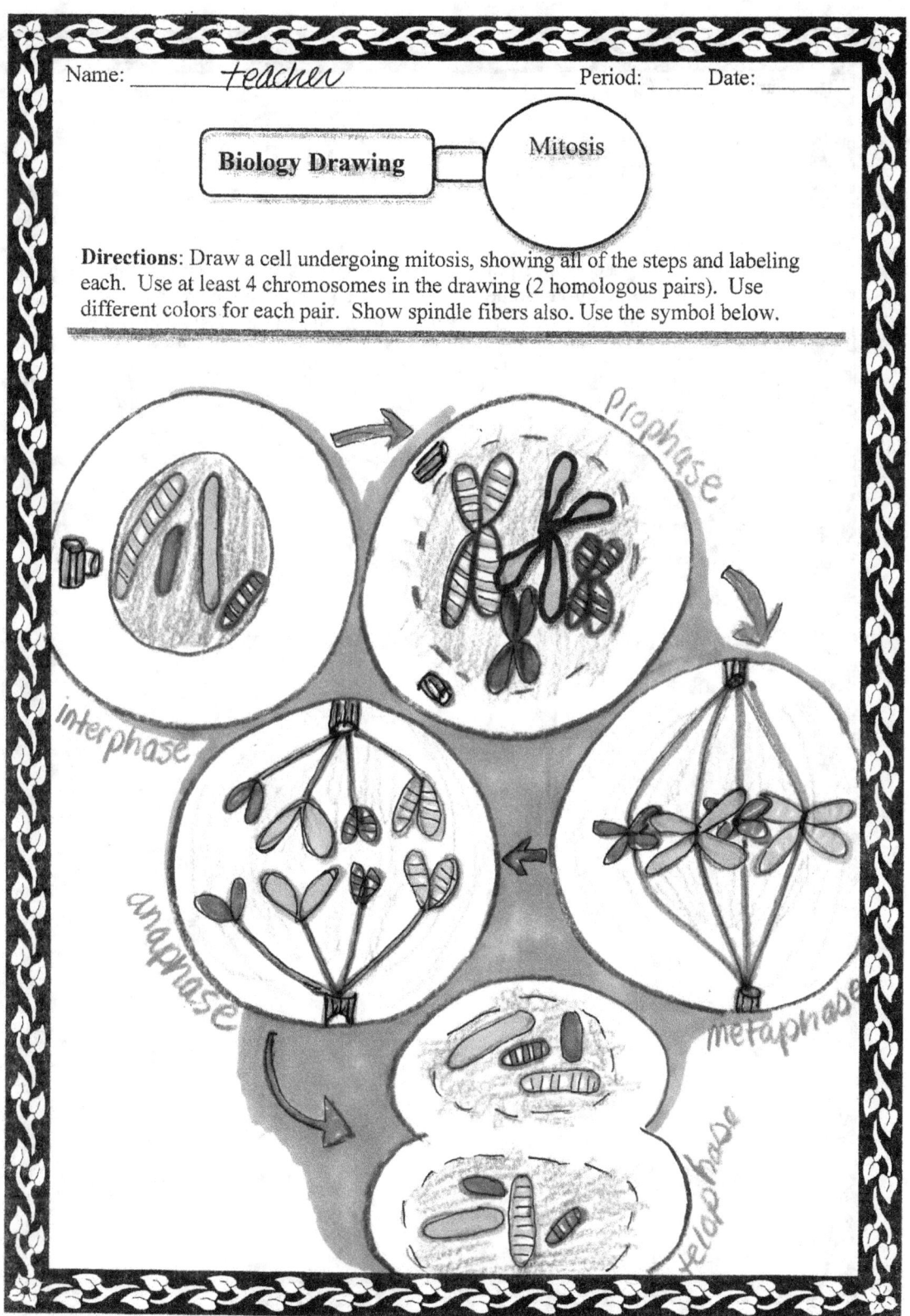

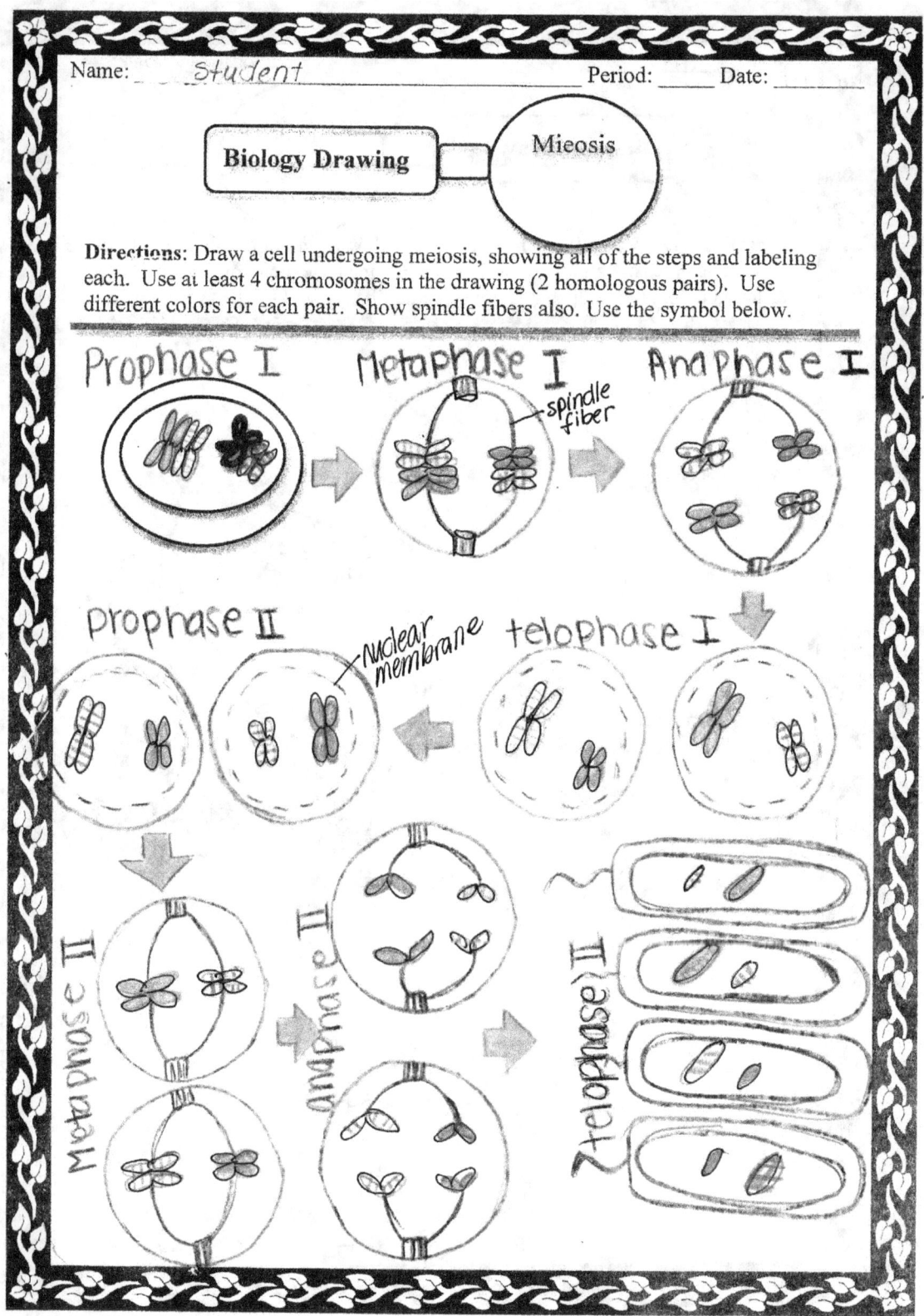

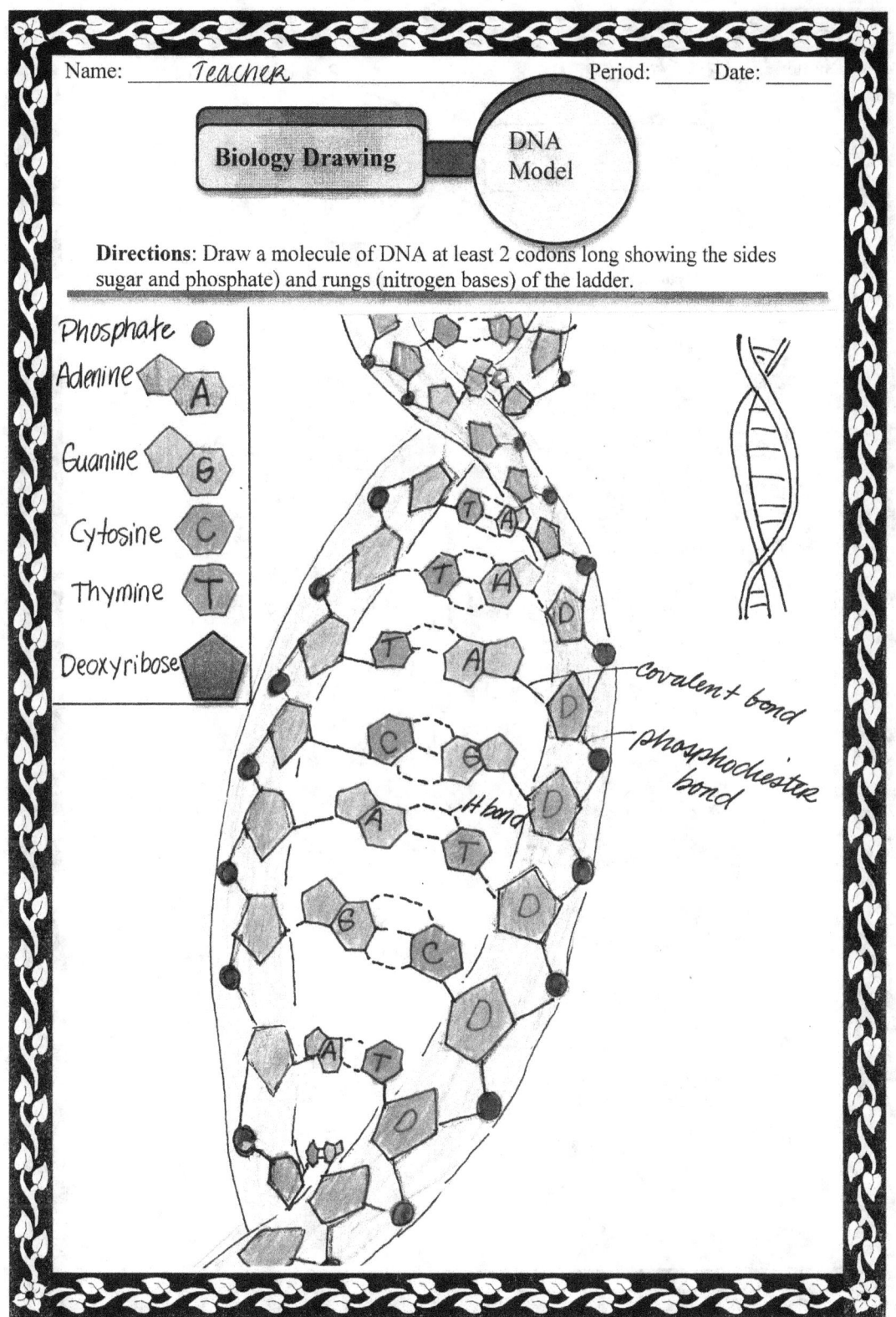

Name: Teacher   Period: ___   Date: ___

Biology Drawing — DNA Replication

Directions: Show how DNA replicates within the nucleus. Use the amount of detail that your teacher requests. Use the symbol provided (can be a nitrogen base).

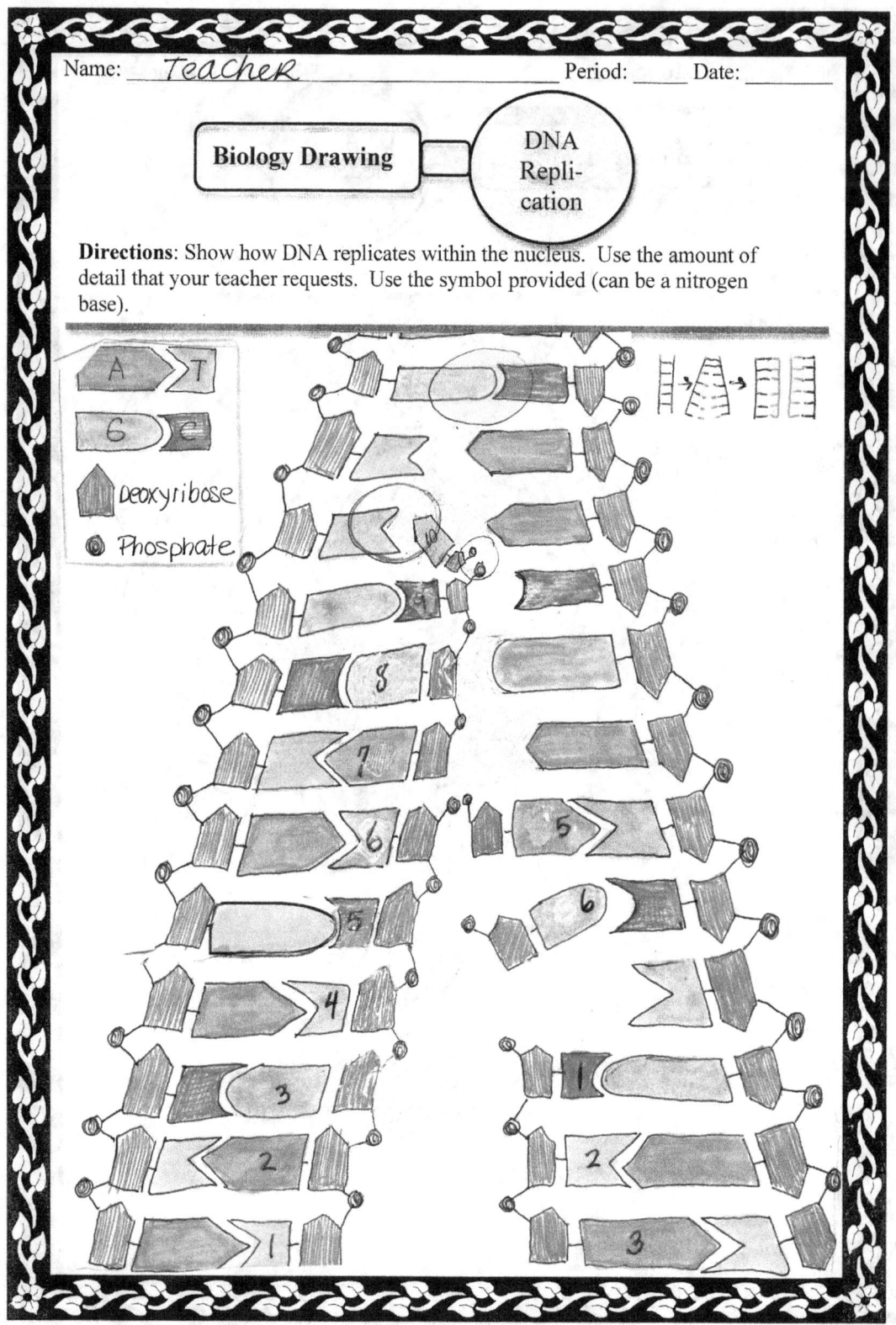

Name: Teacher   Period: Date:

Biology Drawing — RNA

Directions: Show the three kinds of RNA in the space below: Transfer RNA, Messenger RNA and Ribosomal RNA. Use the symbol below in your drawing.

tRNA carries amino acids to the ribosome
mRNA comes from the nucleus where it "transcribed" DNA
rRNA is found within the ribosome

Name: _student_   Period: ____ Date: ____

Biology Drawing — Protein synthesis

Directions: Draw a diagram showing both of the major steps in protein synthesis: transcription and translation. Use the ribosome below in your drawing

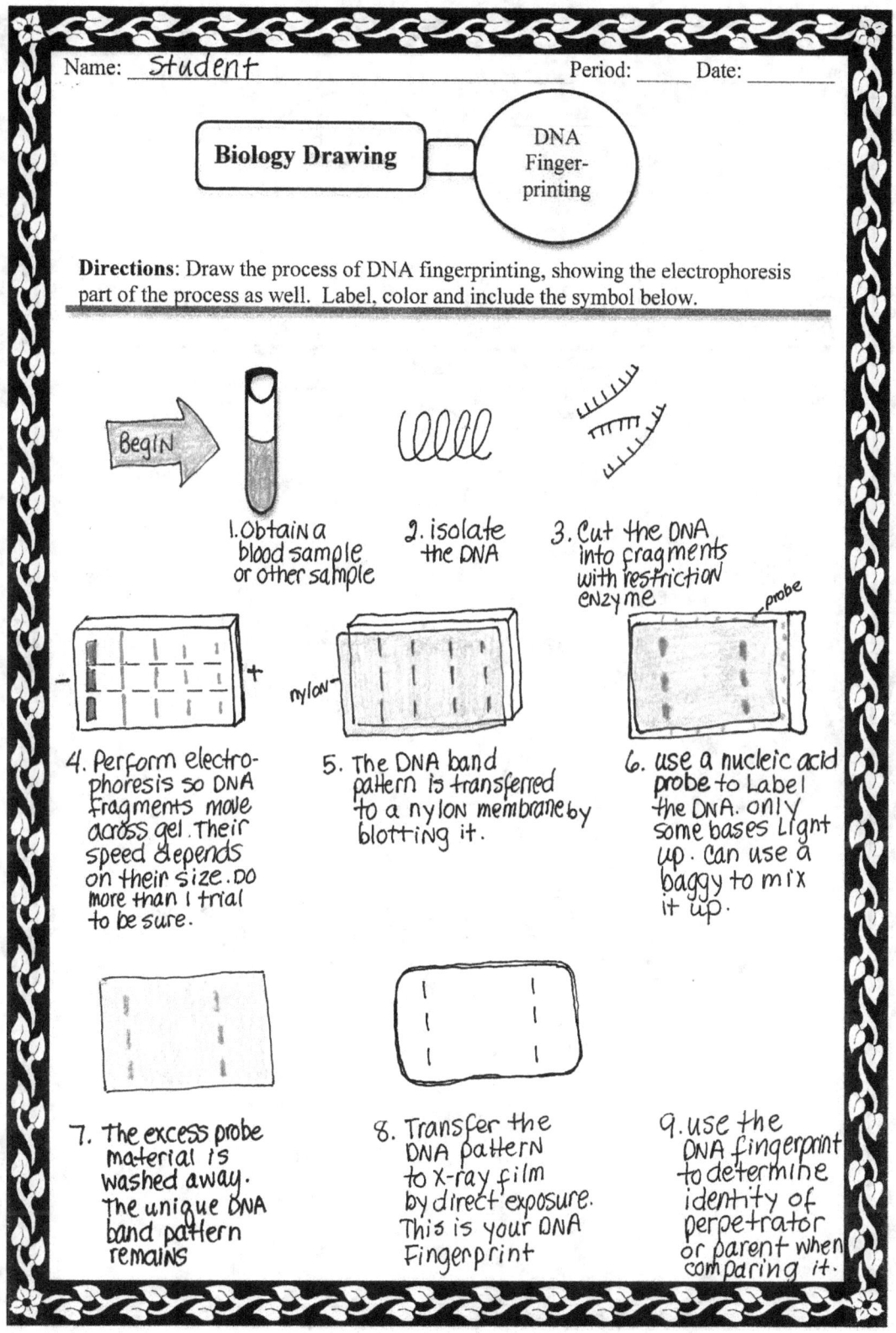

Name: _____ Period: _____ Date: _____

Biology Drawing — Recombinant DNA

Directions: Draw the process whereby recombinant DNA is utilized to make important biological molecules. Be sure to include the plasmid, chimera and donor DNA. Color, label and include the figure below.

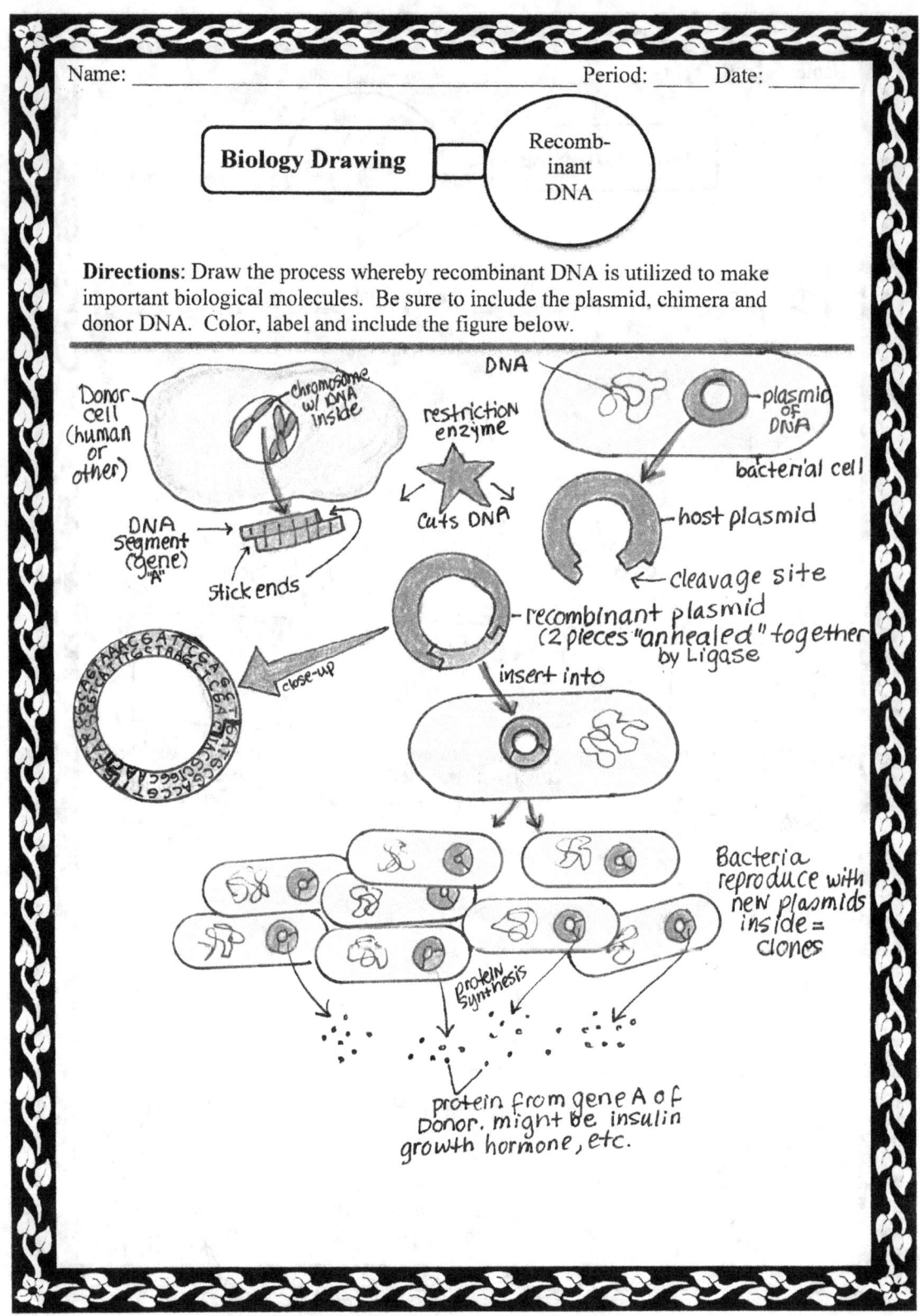

Name: teacher   Period: ____ Date: _____

Biology Drawing — Human Mutations

Directions: Draw a representation of the types of mutations with examples. It can be in a chart form if you like. Include gene mutations and chromosome mutations.

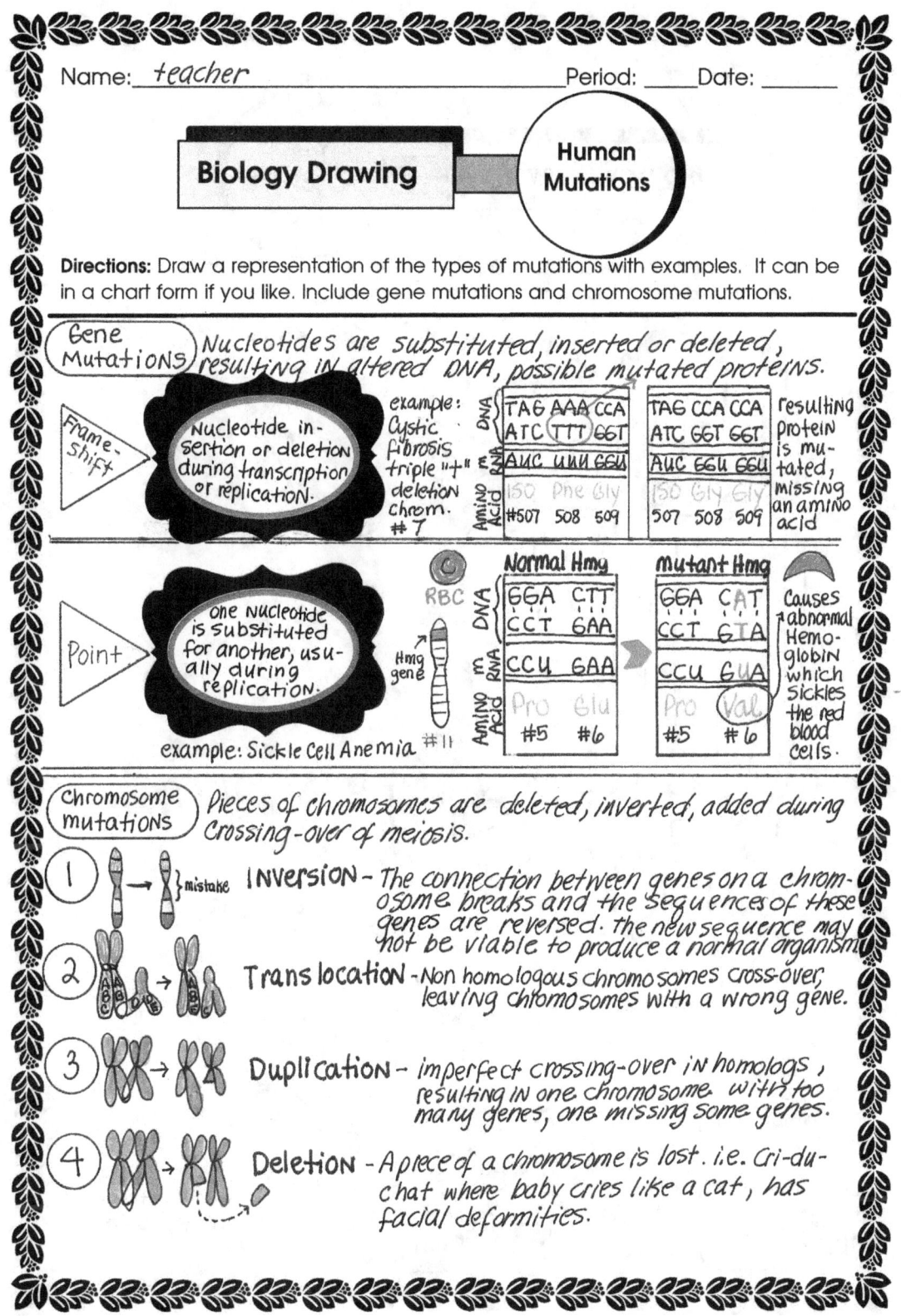

Gene Mutations: Nucleotides are substituted, inserted or deleted, resulting in altered DNA, possible mutated proteins.

Frame-shift: Nucleotide insertion or deletion during transcription or replication.
example: Cystic fibrosis triple "+" deletion chrom. #7

DNA	TAG AAA CCA	TAG CCA CCA
ATC TTT GGT	ATC GGT GGT
RNA	AUC UUU GGU	AUC GGU GGU
Amino Acid	Ile Phe Gly	Ile Gly Gly
#507 508 509	507 508 509

resulting protein is mutated, missing an amino acid

Point: One nucleotide is substituted for another, usually during replication.
example: Sickle Cell Anemia — Hmg gene #11

	Normal Hmg	Mutant Hmg
DNA	GGA CTT / CCT GAA	GGA CAT / CCT GTA
RNA	CCU GAA	CCU GUA
Amino Acid	Pro Glu / #5 #6	Pro Val / #5 #6

Causes abnormal Hemoglobin which sickles the red blood cells.

Chromosome mutations: Pieces of chromosomes are deleted, inverted, added during crossing-over of meiosis.

1. **Inversion** — The connection between genes on a chromosome breaks and the sequences of these genes are reversed. The new sequence may not be viable to produce a normal organism.

2. **Translocation** — Non homologous chromosomes cross-over, leaving chromosomes with a wrong gene.

3. **Duplication** — Imperfect crossing-over in homologs, resulting in one chromosome with too many genes, one missing some genes.

4. **Deletion** — A piece of a chromosome is lost. i.e. Cri-du-chat where baby cries like a cat, has facial deformities.

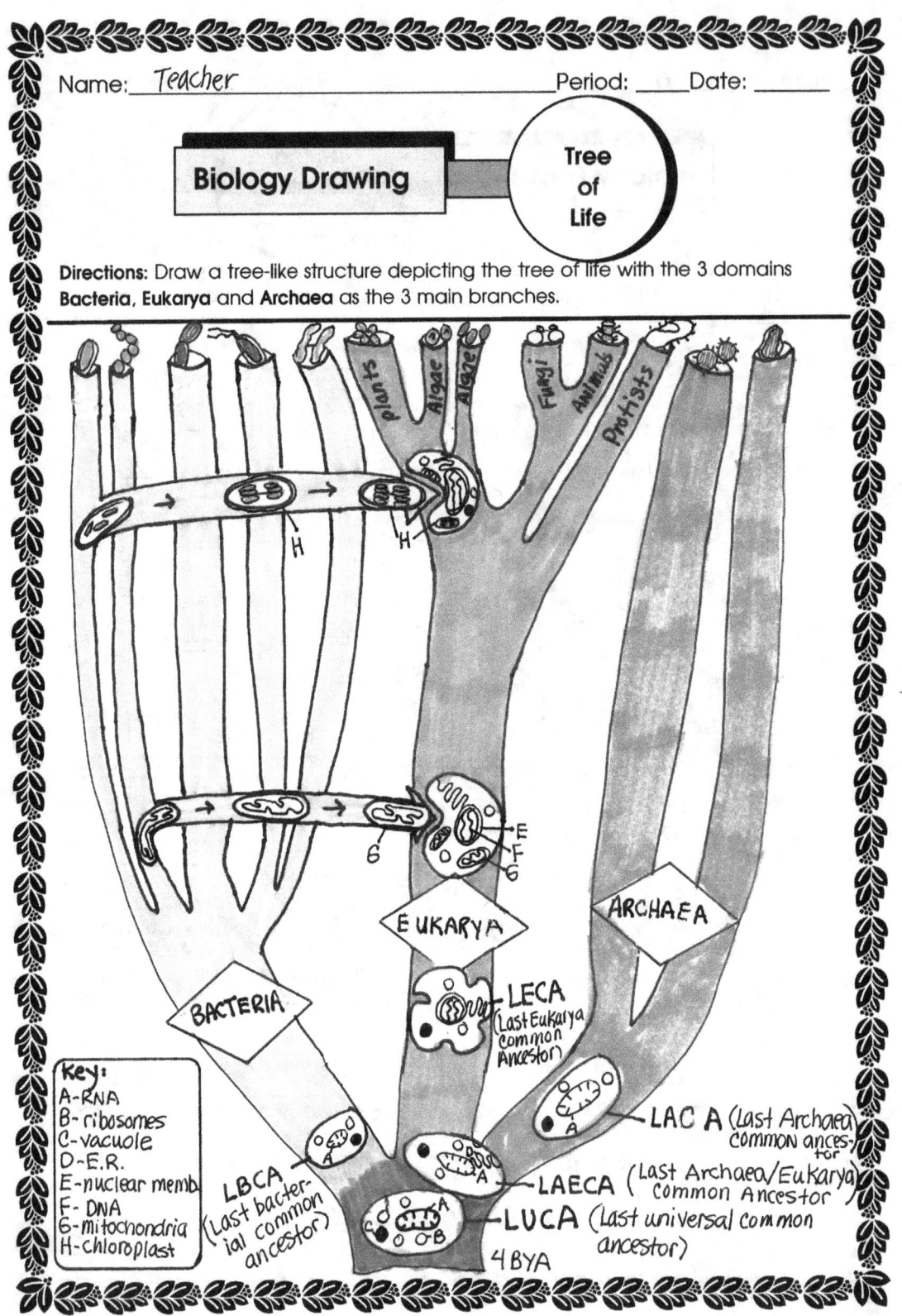

Name: teacher   Period: ___ Date: ___

Biology Drawing — Evidence for Evolution

Directions: Draw a representation of the 5 evidence categories for Evolution, such as fossil evidence, comparative anatomy, comparative embryology, comparative biochemistry and biogeography. You can make a chart if you like.

Type of Evidence	How it gives evidence to evolution	examples
Fossil	Fossil remains show that species are not fixed but can evolve into other species over time. Also, fossil remains have been found of extinct species transitional between major groups of organisms. Use our eyes to interpret.	1. whale - used to live on land. Now have vestigial hip. 2. horse - 60 mya had 4 toes, smaller. 3. elephant - tusks got longer b/c trees got taller. Cools him off.
Comparative Anatomy	The closer a species is to another species, the closer is their _anatomy_. The arm bones and leg bones of a human are more similar to an ape than to a horse.	1. **homologous** structures - same structure, dif function. i.e. horse leg, bat wing, whale fin → adaptive radiation. 2. **analogous** structures - same function, dif structure. i.e. butterfly/bird wing.
Comparative Embryology	The closer a species is to another species, the closer is their _embryological development_. Compare the human's development to other animals, find it is most similar to ape. E. Haeckle "ontogeny recapitulates phylogeny"	Follow development of different animals, see that the closer an animal is to a human phylogenetically, the closer their embryos develop. Phylogeny = development thru time
Comparative Biochemistry	The closer a species is to another species, the closer is there _biochemistry_. You can compare DNA, enzymes, proteins and will see that closer phylogeny shows closer relationships. We all have same DNA code.	1. DNA - the chimp is around 98.77% similar to humans in nucleotides. Same code. 2. hemoglobin - chimp has the same thing as we do; the gorilla has 1 different amino acid!
Biogeography	Organisms in isolated environments evolve differently from each other so that isolation of related organisms over time in different environments results in the development of new species. 175 mya	1. Darwin's finches - geographical isolation → adaptive radiation. 2. pangaea - Finding fossils in different parts of the world where land used to touch tells us of fossils' age.

Name: Teacher   Period: ___   Date: ___

Biology Drawing — Natural Selection

Directions: Divide the page into 5 boxes and show the 5 steps for natural selection as told to you by your teacher. Use the line below to guide you.

Explanation:
1. 100% of population is green
2. mutations occur = variation
3. A change in the environment occurs.
4. Natural Selection occurs
5. 100% of population is brown

Name: Student   Period: Date:

Biology Drawing — **Adaptation**

Directions: Adaptations help organisms to survive and/or reproduce. They can be structural, physiological or behavioral. Pick an animal with a particular adaptation and draw him showing that adaptation in the middle box. In the first box, show what that animal may have looked like millions of years ago before he developed the adaptation. In the last box, show what the animal may look like in a million years with a new adaptation or a change to the original adaptation. Explain what happened on the back and include the symbol below.

Tomarctus 20 mya | GRAY WOLF | WOLF MAN of future

Name: Student   Period: ___ Date: ___

Biology Drawing — Ecology in Levels

Directions: Choose an organism, and then show how it fits into a population, community and ecosystem. Use the circles below.

Name: Student   Period: ____ Date: _____

Biology Drawing — **Species Interactions**

Directions: Draw a "Wanted" poster that depicts any symbiotic relationship of your choice, such as parasitism, mutualism, commensalism, etc.

Name: student   Period: Date:

Biology Drawing — Biogeo-Chemical cycles

Directions: Combine the 3 types of cycles (water, carbon and nitrogen) in one picture. Label, and be sure to write the names of important molecules like carbon dioxide, water, etc. Be sure to include the symbol.

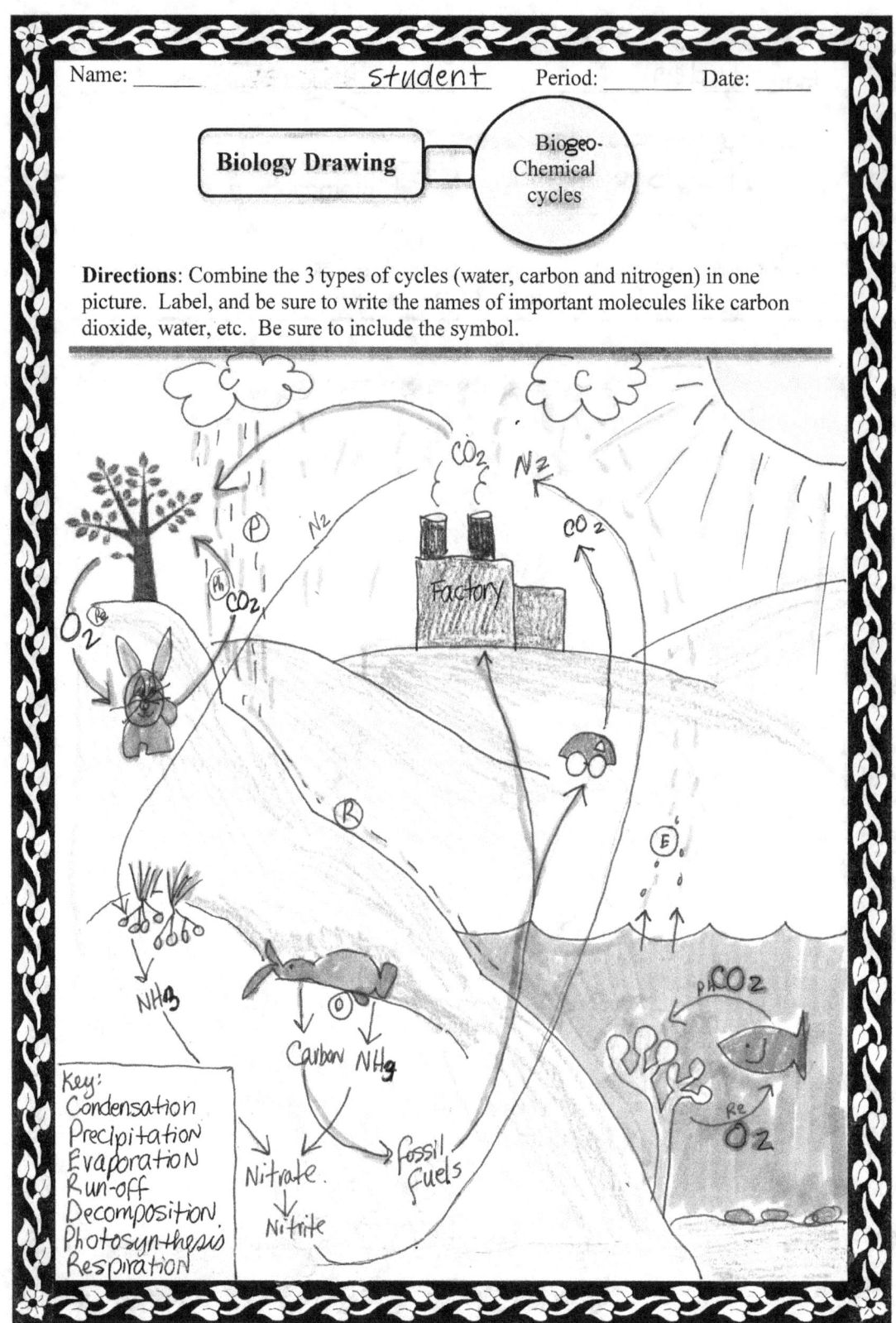

Name: Student   Period: _____ Date: _____

Biology Drawing — Greenhouse Effect

Directions: Draw the greenhouse effect using the sun and ground below. Be sure to show the greenhouse gases, sun's rays, etc.

Name: Student Period: Date:

Biology Drawing — Reduce, Reuse, and Recycle

Directions: Make an advertisement to encourage people to reduce, reuse and recycle. You can choose just one or all three.

Reduce, Reuse, and Recycle your BOOKS!!

Name: _Teacher_ Period: ____ Date: ____

Biology Drawing — Human Body Anatomy

Directions: In the outline below, draw as many major human body organs as you can in the correct position and proportion. Be sure to label the organs.

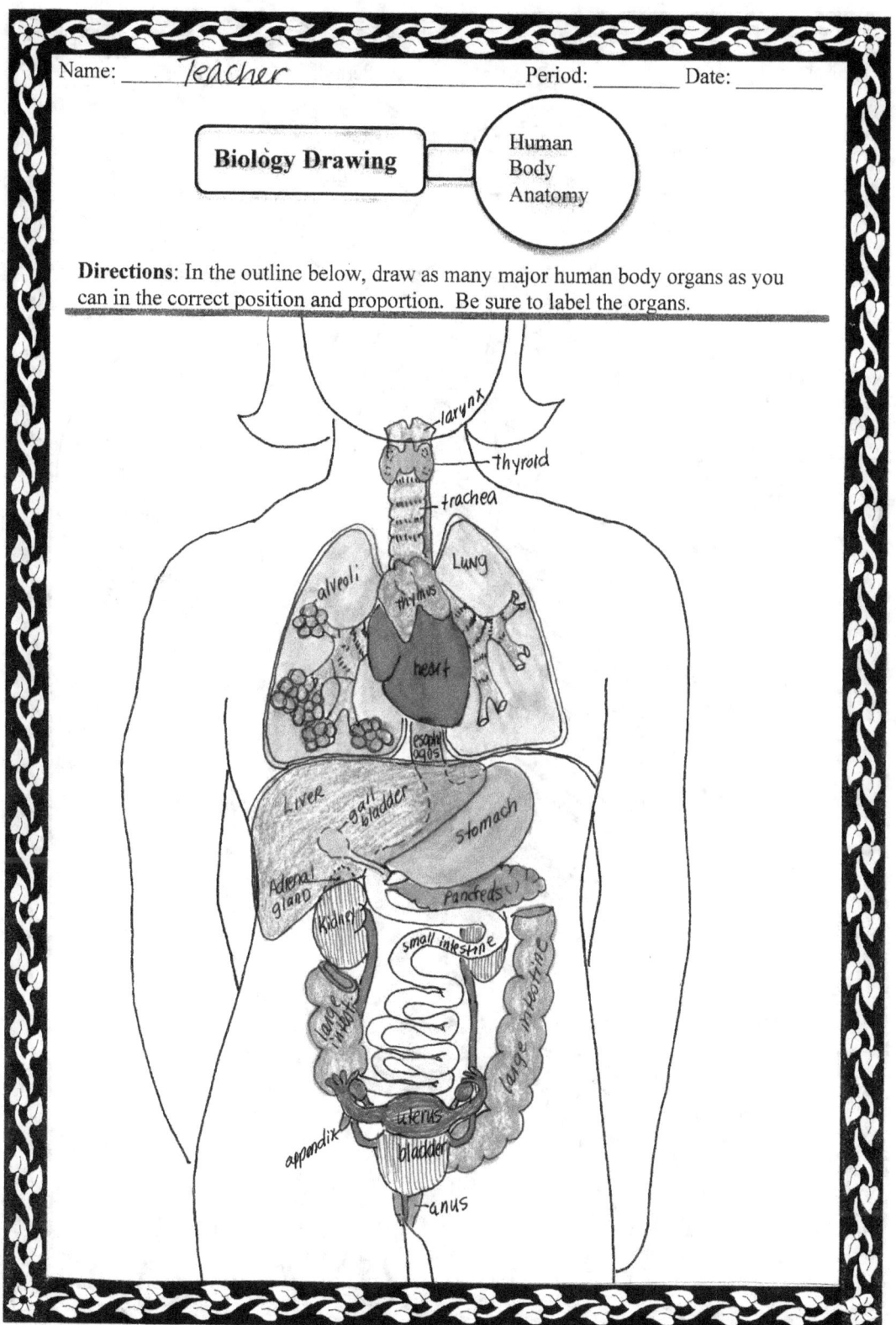

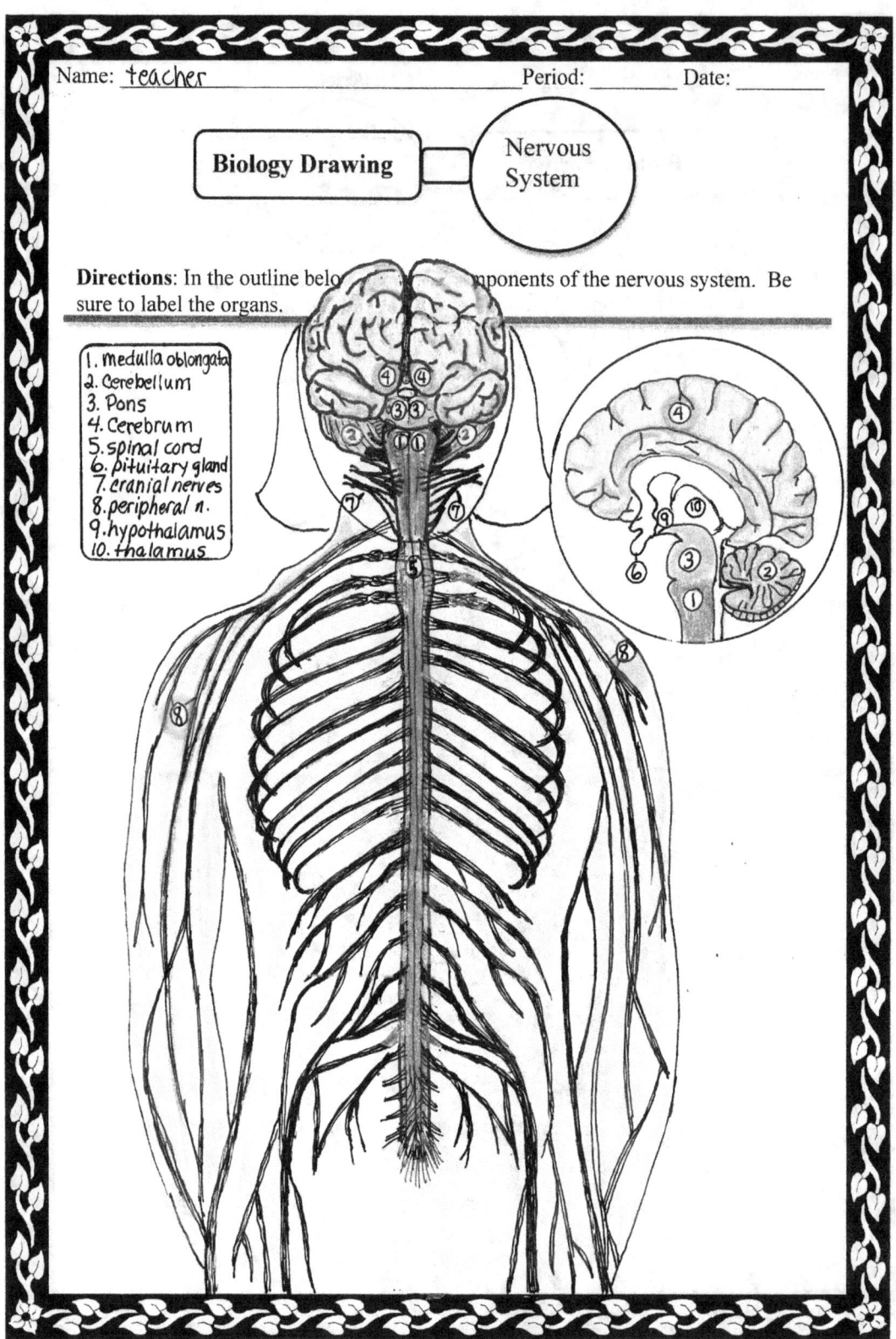

Name: _teacher_   Period: _____   Date: _____

Biology Drawing — Endocrine System

Directions: In the outline below, draw the components of the endocrine system. Be sure to label the organs.

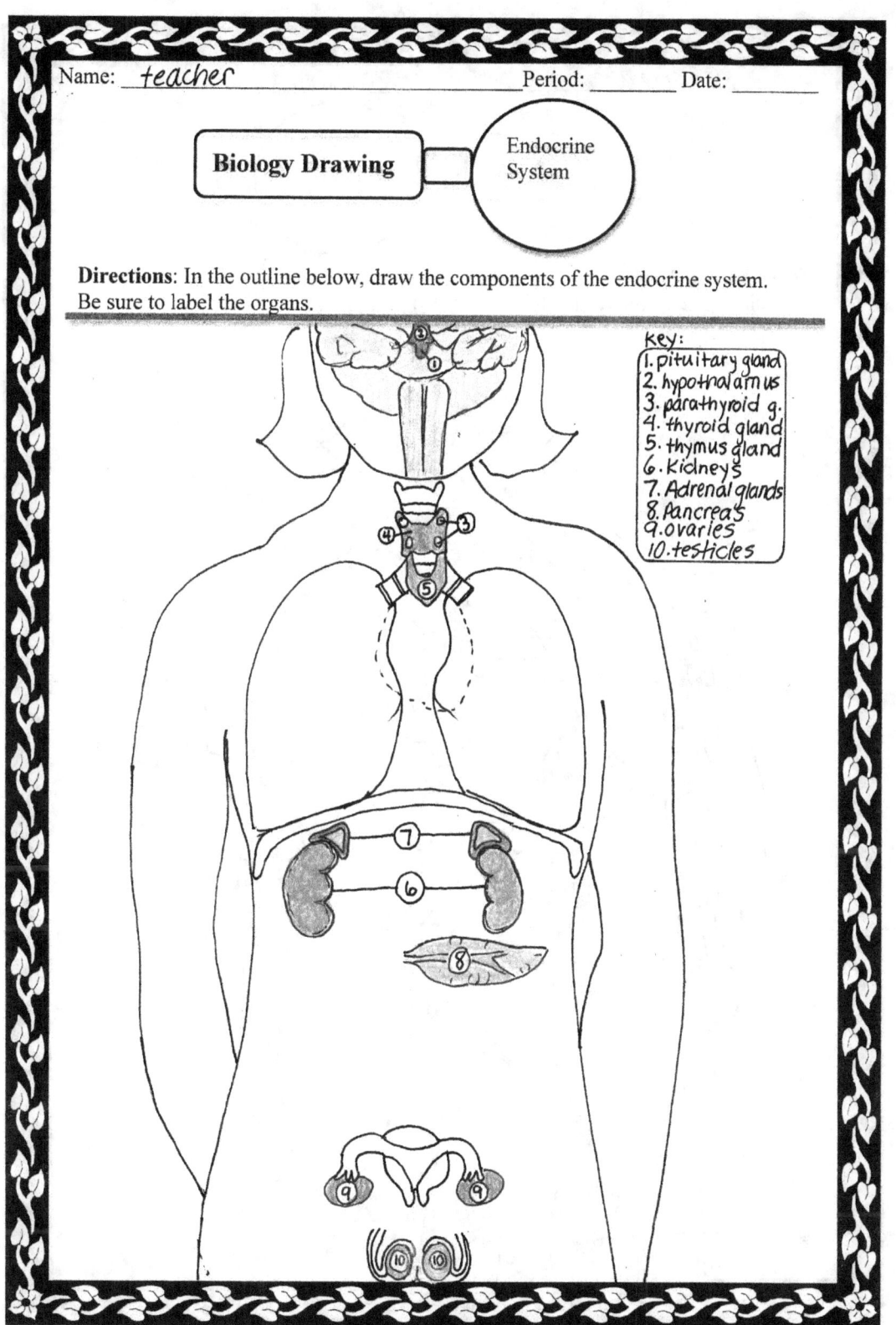

key:
1. pituitary gland
2. hypothalamus
3. parathyroid g.
4. thyroid gland
5. thymus gland
6. kidneys
7. Adrenal glands
8. Pancreas
9. ovaries
10. testicles

Name: teacher   Period: Date:

Biology Drawing — Immune System

Directions: In the outline below, draw the components of the endocrine system, including glands, organs, etc. Be sure to label the organs.

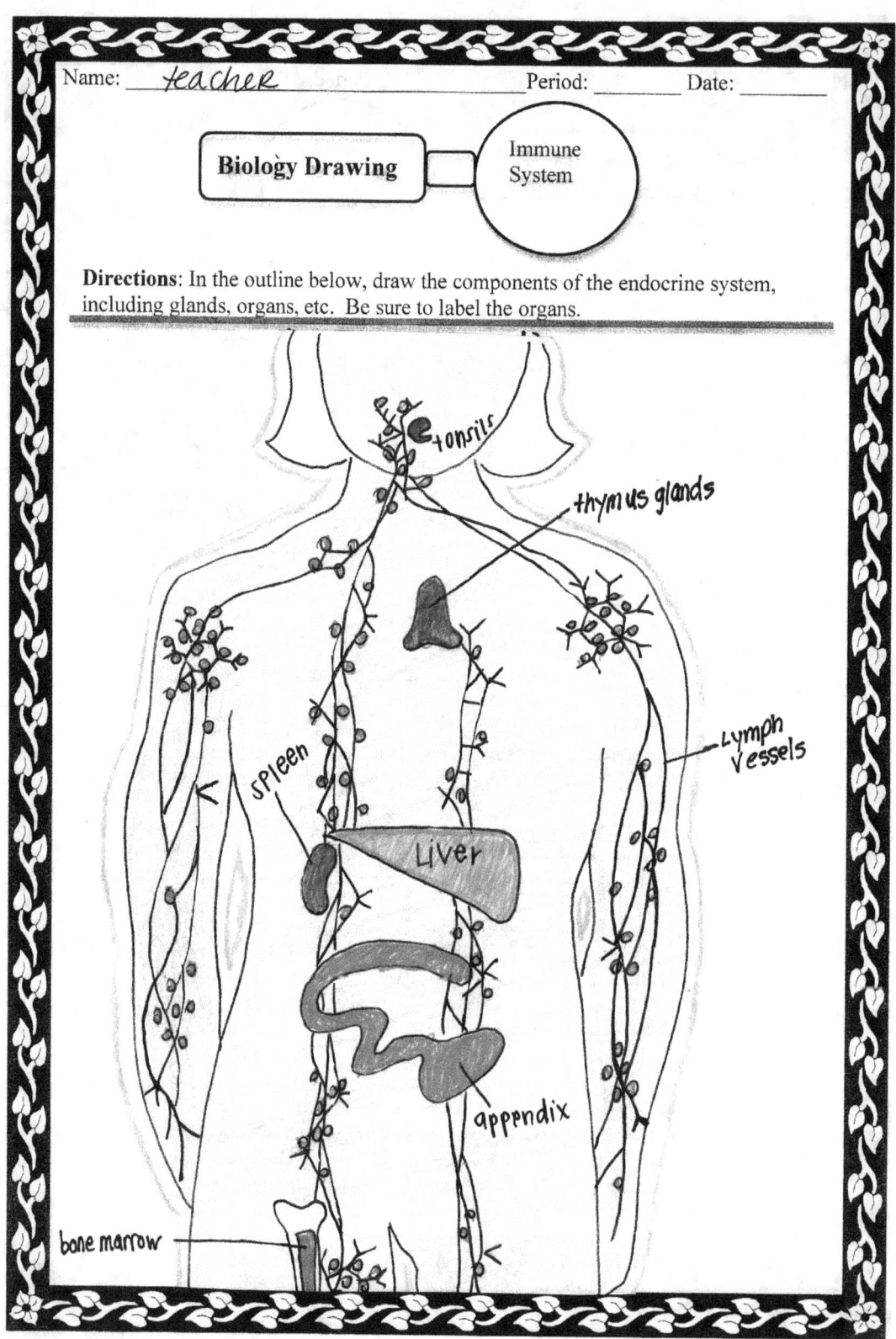

tonsils
thymus glands
lymph vessels
spleen
liver
appendix
bone marrow

Name: teacher   Period: ___   Date: ___

Biology Drawing — Integumentary System

Directions: Draw a section of the human skin, showing the different layers and important structures. Use the line below in your drawing.

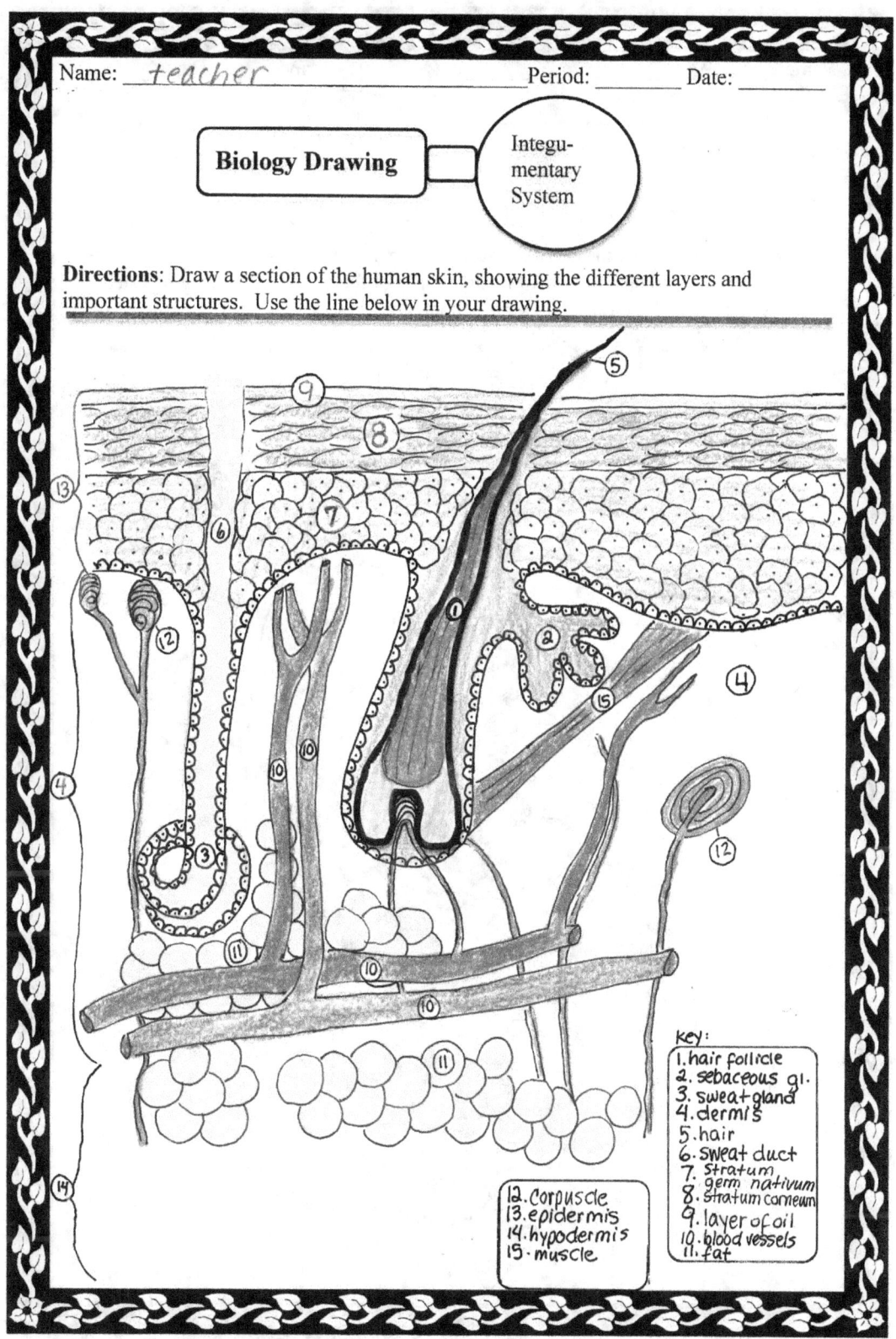

Key:
1. hair follicle
2. sebaceous gl.
3. sweat gland
4. dermis
5. hair
6. sweat duct
7. stratum germ nativum
8. stratum corneum
9. layer of oil
10. blood vessels
11. fat
12. corpuscle
13. epidermis
14. hypodermis
15. muscle

Name: Teacher   Period: Date:

Biology Drawing — Digestive System

Directions: In the outline below, draw all of the digestive organs, label and color. Then, if directed to, show the three types of food molecules being digested along the way. Use circles for carbohydrate monomers, squares for proteins (amino acids) and triangles for fats.

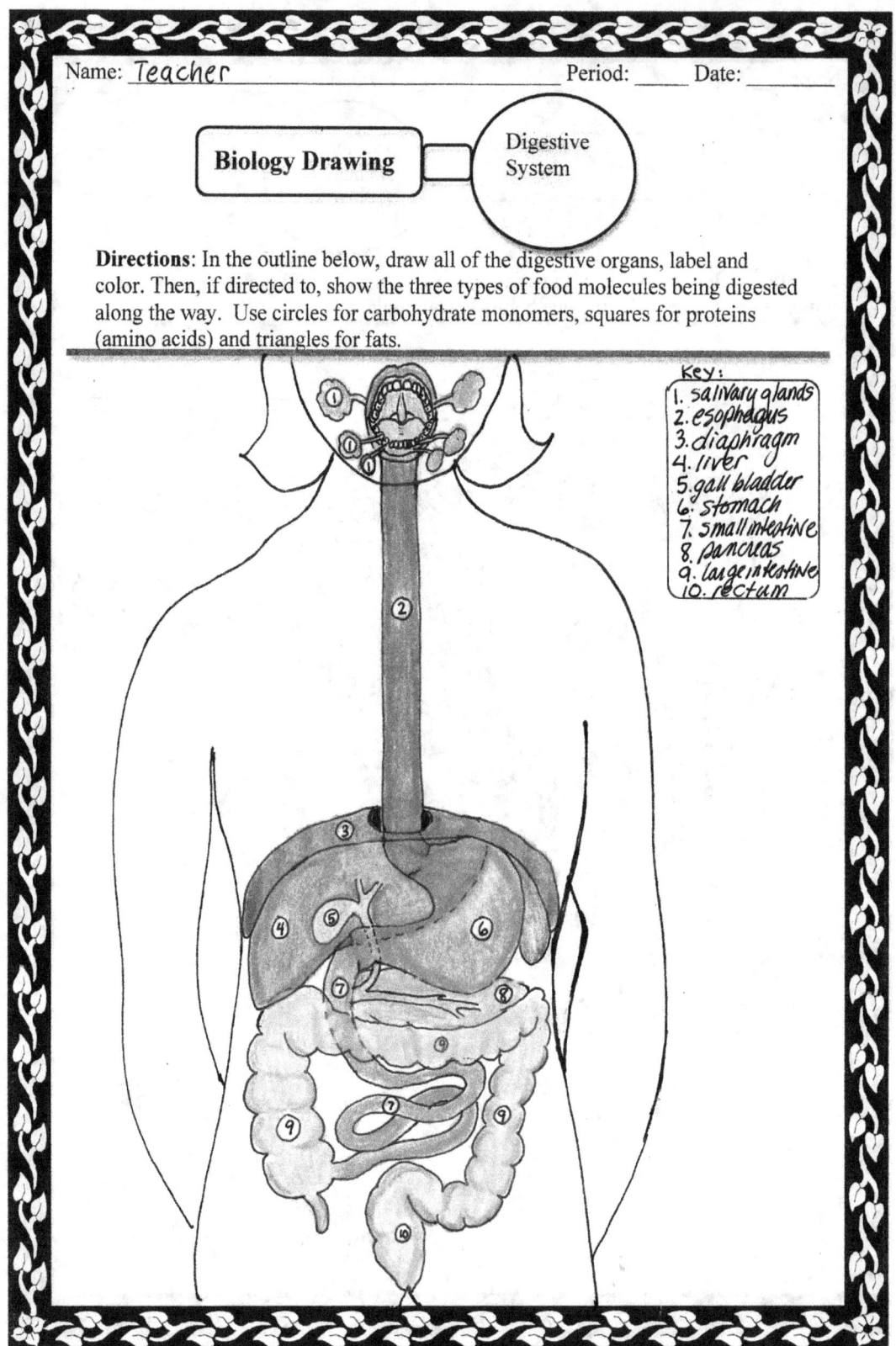

Key:
1. salivary glands
2. esophagus
3. diaphragm
4. liver
5. gall bladder
6. stomach
7. small intestine
8. pancreas
9. large intestine
10. rectum

Name: Teacher   Period: _____   Date: _____

Biology Drawing — Choose a Body Organ

Directions: Draw a body organ of your choice showing as much detail as required by your teacher. Be sure to label and color. On the other side, explain where the organ is found in the body, what it's job is, and whether it is essential to body function. Use the symbol below in the drawing.

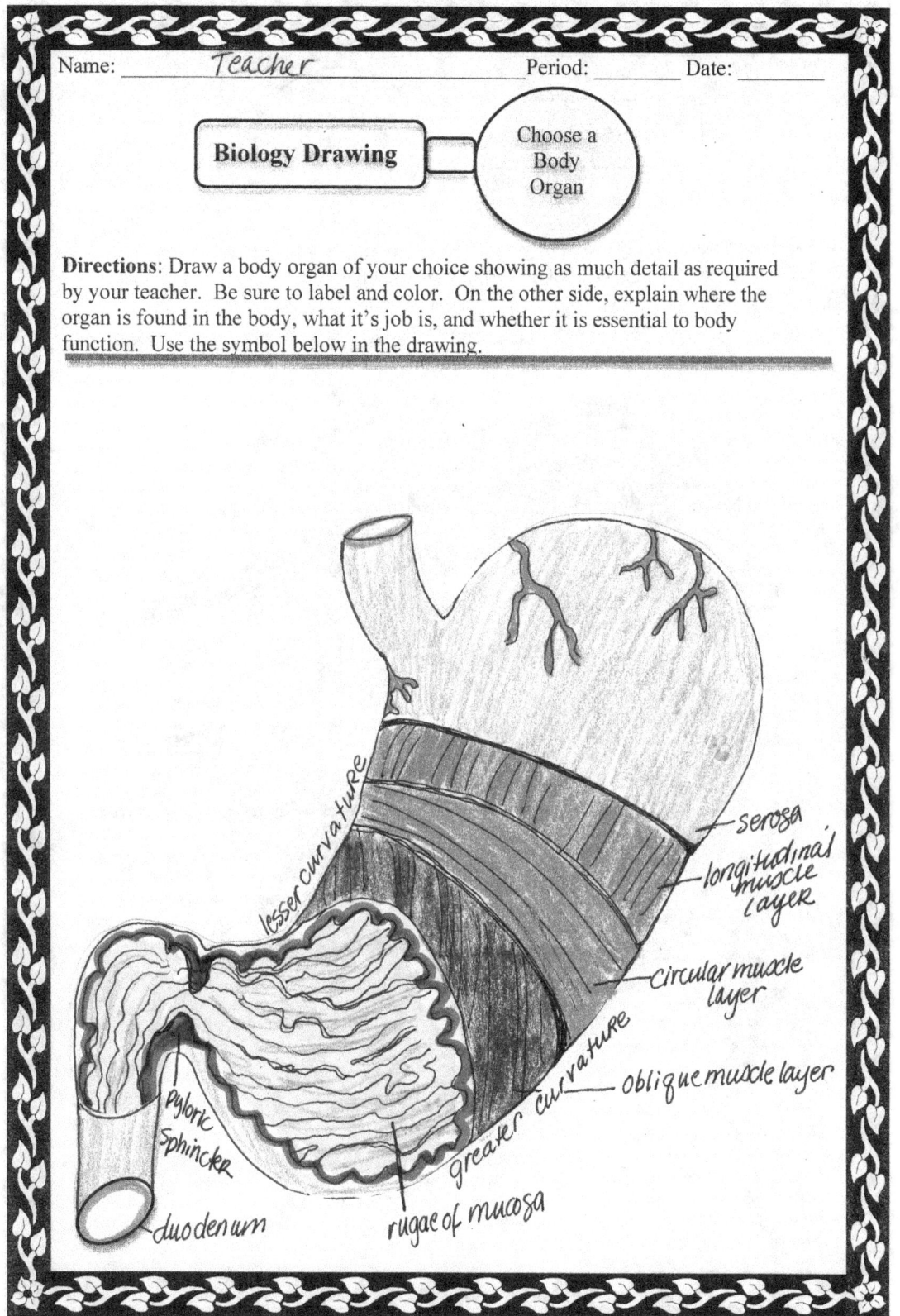

Name: Student   Period: Date:

Biology Drawing — Choose a Body Organ

Directions: Draw a body organ of your choice showing as much detail as required by your teacher. Be sure to label and color. On the other side, explain where the organ is found in the body, what it's job is, and whether it is essential to body function. Use the symbol below in the drawing.

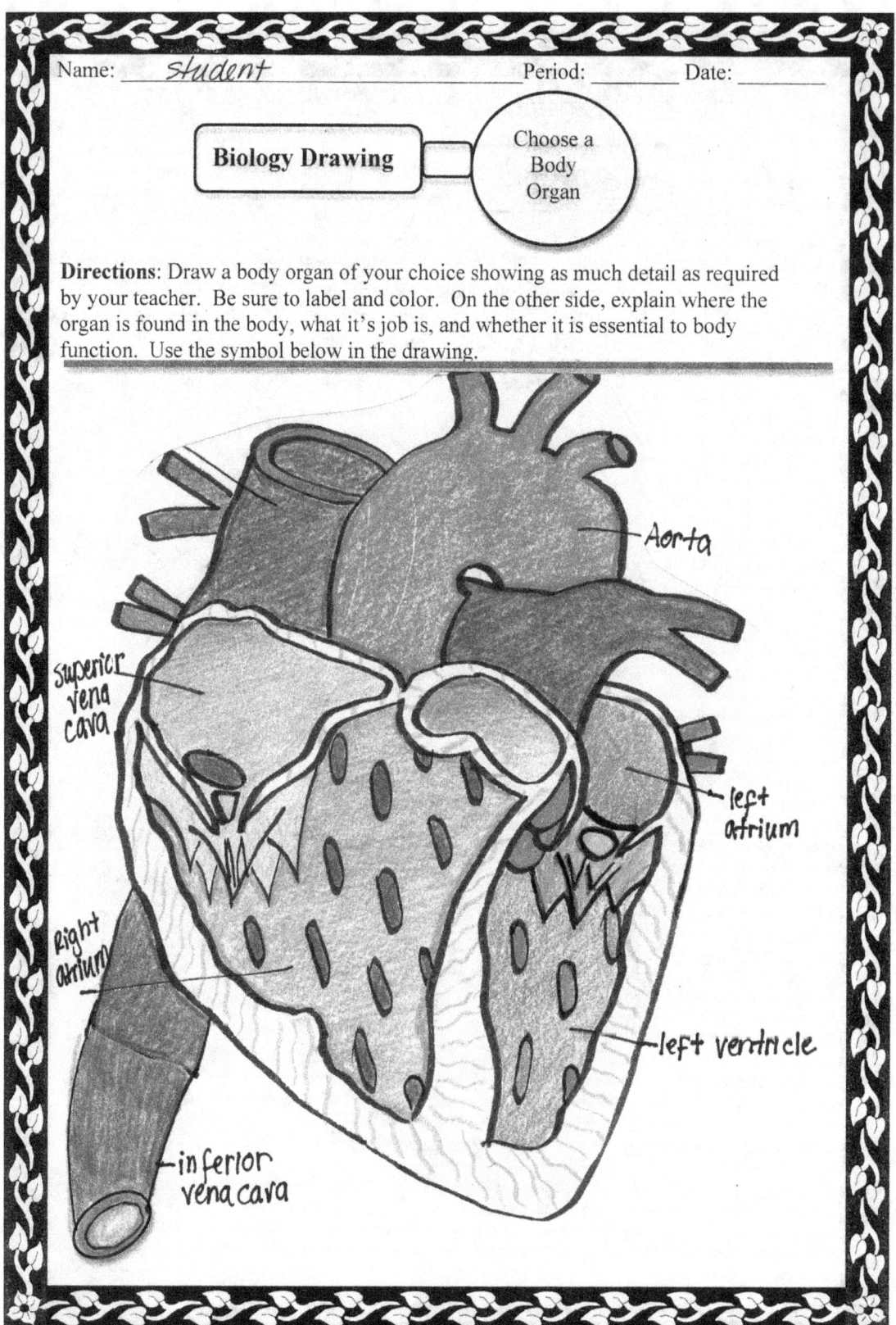

Name: _Teacher_   Period: ____ Date: ____

Biology Drawing — Anatomy of the Frog

Directions: Draw a diagram of the frog, showing it's internal anatomy. Be sure to label and color.

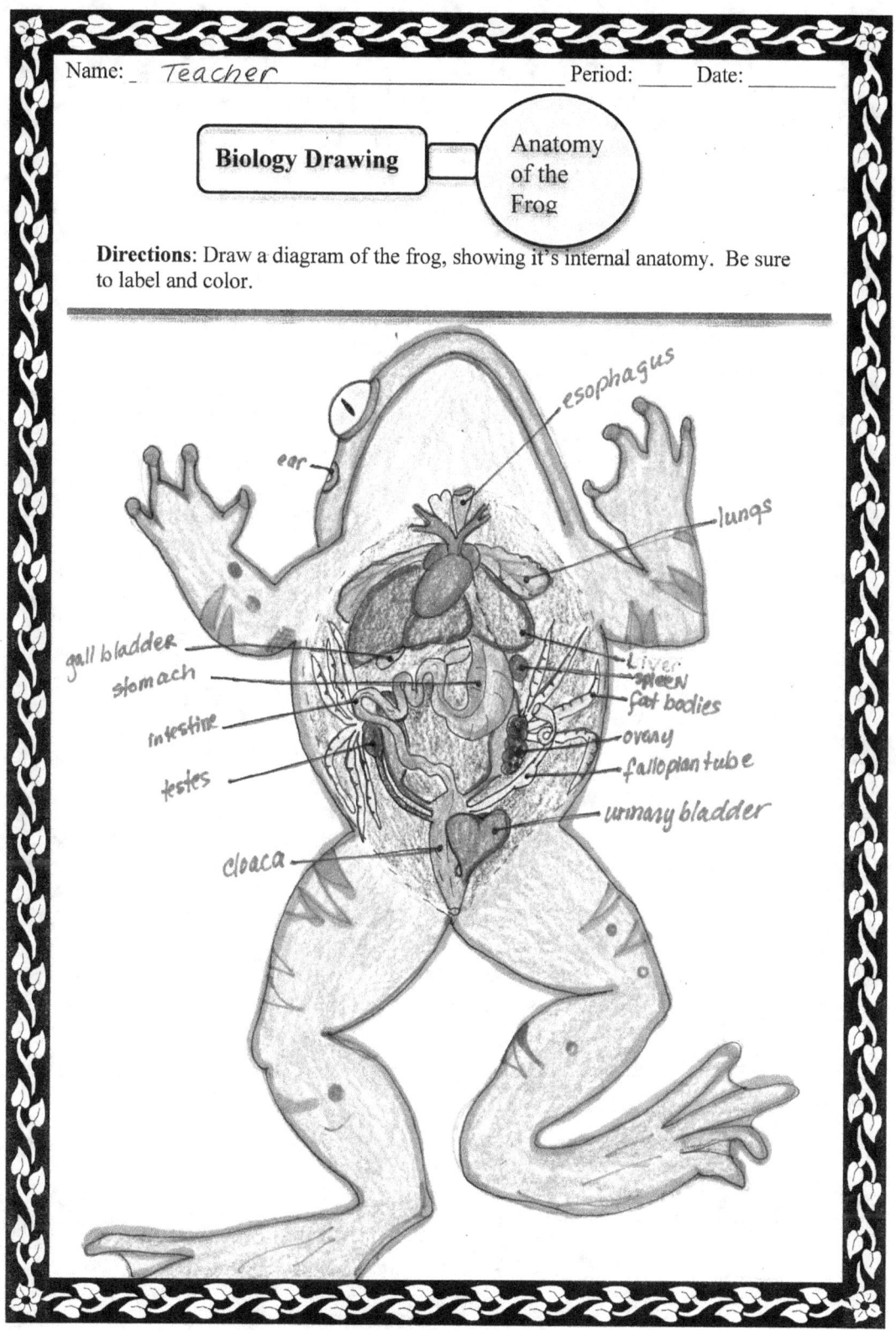

www.ingramcontent.com/pod-product-compliance
Lightning Source LLC
Chambersburg PA
CBHW081151180526
45170CB00006B/2025